SIX DECADES: 1955 to 2015

Introduction

In December 1955 I turned 7. In December 1955 I had just started elementary schooling. This was the year of my awakening. The World suddenly opened up; it extended outside the immediate family circle and beyond the playground of the pre-school. I realized, and was wonderfully shocked, that Life had started much earlier than yesterday, way before I was born, and that it will continue indefinitely after my death. I was seven; the age of awareness, that for every happy birth there is a sad death. However, the why, the before and the after were still foggy concepts. One thing for sure, life was a very long trip with many routes and one destination; my route was not yet charted.

Walid Nasr

Dedication

I dedicate this book to:

My eldest son whose early departure gave me the ultimate test of endurance; forgive me Marwan for the many times I failed this test;

My daughter Mira who, notwithstanding her own lovely family and prosperous career, will always be my little girl;

My youngest son Majed whose maturity and good heartedness are far beyond his young age;

My grandchildren, Kiira, Kaila and Alexander, who brighten my life and give a meaning to joy; and,

My wife Vicky who is simply my life on earth and beyond.

Acknowledgement

Many thanks to my Father who, from my early youth, instilled in me the habit of reading, for joy as well as for mental nourishment. To Khalid Racy, who was my mentor during my first year of work; thank you for insisting that however busy my work schedule is, there should always be time for reading. Natheer, during our few years together with the UNDP in Kuwait, you were genuinely interested in our continuous discussions on a wide spectrum of subjects; I followed your sincere advice and wrote my book.

My gratitude to my community of Monsefites, my friends and companions during these Six Decades and my acquaintances who gave wealth to this book; you have all enriched my life.

Adonis, our friendship is endless; you are part of every memory I have, of the days when selfless friendship and loyalty meant something.

GOD, I humbly thank you for blessing me with a family that turned every house we lived in, and they are many, into a home full of gratitude and happiness.

Author's Note

I started this book as an introduction to management; especially project management. The more I wrote the less excited I got; I was simply bored. It dawned on me that I was repeating what many others had already written and with more theories, graphs and factual examples of real cases.

I decided on another approach; management as I experienced and practiced during 45 years of diversified work environments, on a multitude of projects and in several managerial positions. Remembering these 45 years took me to several countries, many cities and different cultures; memories of pleasant and not so pleasant encounters, happy and sad moments. Management concepts became intertwined with the facts of my life, with occurrences that I witnessed, with people who played a role in creating my memory bank, and mostly with what went on or how I conceived of things around me.

My book now took the shape that I felt well with; it is not an autobiography but a panorama of the World that I lived in over "SIX DECADES: 1955 to 2015".

The facts in the book are as I remember them; any distortions or misinterpretations are not intended.

The people who are named in the book are real and appear in the context of my interaction with them; I have used First names only to maintain anonymity to the extent possible.

The management concepts that are defined and described in the book, italicized and copied in Appendix I, reflect my experiences and practices; they should not infringe on the rights, thoughts and theories of authors of management concepts.

Finally, my sincere apologies to those who believe I have wrongly interpreted facts; to the people I named, if I have in any way offended them; and to experts who might believe that I have oversimplified management concepts.

SIX DECADES: 1955 to 2015

Prologue

AND HUMAN LIFE GOES ON. It is like a train in perpetual motion advancing on a rail that moves only towards the future. The cornerstone of this rail was placed by God or the Darwinian Theory or some other supernatural or coincidental incident, at an unknown moment in time. Religious revelations and scientific discoveries are relentless in their efforts to determine the "when" of creation. The railway has an ever increasing number of side tracks and train stops that grow with time; but none can stop life's forward motion. The development in global communication and in statistical data analysis can pretty much predict the number of people getting on or off that train, at the birth platform and the death platform of each station. I got on the life train on the fourteenth day of December of the year 1948, a date I had no control over, and shall eventually get off at the stop of my final destination. The length of my trip is beyond my control as long as I do not jump off before reaching my station. The richness and the varieties of the panoramas running past my window; the number of passengers I interacted with; and the multitude of activities I participated in cannot be fully and justly described during the remaining leg of my trip. My sincere hope is for my memory not to fail me or to deceive me in my attempt to share these scenes with my family and project the happenings of my trip to the readers.

CHAPTER I: 1955 to 1971

Engineering Class of 1971

He stood tall among men, Dr. Raymond Ghosn, the Dean of the School of Engineering and Architecture of the American University of Beirut, better known as AUB, proud to be the humble motivator of a new generation of professional engineers. He cleared his throat and delivered a most shocking message to us, we the superhuman engineering graduates of 1971:

"Congratulations graduates of 1971. You are now equipped with the ability to start learning the profession of electrical, mechanical, civil engineering and architecture. When you go into the professional arena of the outside world, you will find out that you know less than anyone in the engineering business; however, I am confident and so should you be that in no time you will surpass those who have preceded you, and God willing you will excel."

With those words, and the euphoria of graduation mixed with the disappointment that we were not going on stage to parade in ceremonial gowns, we walked out of the assembly hall.

The year 1971 had started with a series of sit-ins and strikes by the AUB students, protesting against Arab weaknesses and Israel's dominance. The Palestinian issue was the overwhelming source of ideologies in the Arab World, and AUB was the breeder of leaders, politicians, and those hoping to make a difference in their Middle Eastern societies. AUB also had its share of those who attached themselves like leaches to sincere political and social activists, hoping for a free ride to the limelight of recognition, and motivated by the pleasures of idleness, laziness, running away from classes, and with luck landing some easy-to-impress chicks. The AUB student body, like most of the Arabs, had not yet recovered from the humiliating defeat during the 1967 Arab-Israeli war. The war was over in very few days, but long enough for the Arab armies to be defeated and the Arabs to realize that their conquests belong to a past kept alive by nostalgic songs, low budget movies, and theaters with parading actors in turbans, colorful robes, and dangling swords. Some carried daggers, mostly used in acts of treachery or treason.

The year before, witnessed a student revolt within AUB that was culminated by a shocking and unprecedented act during the graduation ceremony. A student, sympathizer to Fatah, the Palestinian political and militant movement, decided to deliver a message, on stage, in support of Pro-Arab cum Palestinian ideology and the growing anti-American sentiments. To do that, he snatched the microphone from the master of ceremony, in not so gentle a way, and managed few angry words before the charade ended. The sensitivity of the AUB management was heightened to a fine level, prompting them to take drastic action in response to the slightest spark of mental or behavioral revolt. To avoid a repeat of the previous year's incident, and maybe to deliver a tough lesson to the student body of AUB, the management's reaction to the sit-ins of 1971 was to cancel the graduation ceremony.

Graduation ceremonies are considered by the Lebanese as reward and payback to the families and relatives of the graduating students for their sacrifice, belt-tightening, and many sleepless nights, in support of their children, financially and emotionally, through their academic years. Most Lebanese families believe that good education is restricted to private schools and universities, and these do not come cheap. In fact, many families have to pool their resources and do without the luxuries as well as the necessities of life to pay the education fees. Financial aid is scarce; and quite often governed by social and political rules not related to academic achievement or need. So, when their sons and daughters parade on the stage, a message is delivered to all attendees of the graduation ceremony: My offspring is intelligent, so am I; we, the parents have suffered so we are good; and hey, we are a family of achievers. The message becomes far more powerful, and increases exponentially, as manifested by the applauding and whistling, when the offspring graduates with honorable mention, distinction, and best of all, high distinction. The achievement of one unlucky student, who had the misfortune of not attending his graduation ceremony at the International College of Beirut, was thought to be the highest ever by few scattered attendees, when the Dean of Students followed the student's name by the phrase: In absentia. That had to be the very best, as only one student received it. Not all Lebanese were educated as well as their children, and maybe those few applauders were French educated.

Education System in Lebanon

The education system in Lebanon was inherited from the French during their mandate after World War Two. It is the Baccalaureate (BAC) system, with BAC-1 which is equivalent to 12th grade in the American system, or the last year of high-school, and BAC-2, an additional one year of high-school, which is equivalent to freshman year, or the first year in an American university. The BAC system consisted of 6 years of elementary education and 7 years of high-school, ending with BAC-2. At one time in the recent history of Lebanon, passing a government obligatory test, the "Certificat", was necessary to move up from elementary to high school. At the end of the 4th high-school year, passing another government test, the "Brevet", was enforced by some schools but not all. However, passing BAC-1 government exam was a pre-requisite to move to seventh high-school year; and passing the BAC-2 government exam was essential to be admitted to the professional schools of the universities in Lebanon, namely Engineering, Medicine, Agriculture, and Law. Those students who wanted to study liberal arts could do that after passing BAC-1 exams without suffering the humiliation of further screening by BAC-2 exams. This also applied to those privileged students who were financially fortunate enough to continue graduate studies in universities abroad.

The education system was also classified as English or French, depending on the language in which the sciences and math subjects were taught at schools. History, geography, and civics were studied in the Arabic language in both systems; Arabic, English, and French languages were also taught in both systems. The advocates of the English system believed that the system paid equal attention to the development of the students' character as well as to their education, and paved their way to better adapt to university life and environment. Advocates of the French system believed that this system focuses on intensive studying and trains students to absorb large amounts of information, which would eventually allow them to excel in American universities as well as in French universities.

In the sixties and seventies, French system students had the choice to study at any of the well-known universities in Lebanon (American University of Beirut (AUB),

Lebanese American University (LAU), Lebanese University, Arab University, Haigazian, University Saint Joseph etc.), or abroad, and major in any of their offered professions. However, students of the English system rarely mastered the French language and thus were limited in their pursuit of higher education to American universities; they had no chance of studying dentistry or law in Lebanon, which were given in French only, at the French University.

The difference between the French and English educational systems goes deeper than that. The French language schools were mostly run by religious missionaries who adopted and enforced a far stricter code of behavior than the more liberal English language schools. The latter were mostly influenced by the American culture, which was spreading like a wild-bush fire, benefiting from the introduction of the Cinema and later the Television. It advocated more freedom of expression, less rigid codes of dress and conduct, and most importantly it allowed the students a limited but highly valued participation in setting up the regulations that defined the relationship between the student body and the school administration. Lebanese families, who equated French language with sophistication and considered French traditions as a reflection of upper class finesse and social refinement, were adamant about sending their children to French language schools. On the other hand, families whose rebellion against the French domination still raged inside them, even-though the French mandate was over; or families who had the foresight to realize that the American culture and the American foreign policy were going to control the rest of the world, even if by force, ended up sending their children to English language schools.

Our Family Education

My father was born in Monsef - Lebanon, in the year 1920. His father, who was also born in Monsef - Lebanon during the reign of the Ottoman Empire, was a medical doctor who graduated from St. Louis State University in Missouri, USA, in 1899. After graduation, and upon his return to his hometown in 1901, he was drafted, like most of the males of his community, in the Ottoman army to serve during World War One at the battle-front. My grandfather survived the war, and returned home to father six children, in addition to the two who were born before he was drafted. My father was the first of the postwar new pack of six, and the first of the only two male children. Growing up in Lebanon after World War One was a hardship he had to endure. The responsibility to help in putting food at the family table and to secure the basic necessities for decent survival, while pursuing his quest for knowledge and learning, when the tuition fees and the books were not available, was not an easy task to say the least. Childhood and its carefree approach to life had to be skipped all together. Yet, he managed to prevail, and graduated from the American University of Beirut with a BA in Economics, with honors, at the age of nineteen. During his earlier school years, at Shwaifat high school and later at the AUB, he associated with those students, Lebanese and of other Arab nationalities, who shared with him the defiance against the French mandate, which to them was a French occupation. So, it was natural for me to end up in the English language school system.

There is a lot of truth in the saying "like father like son", especially when the father consciously influences his son to be like him. Parents' influence could be by direct or indirect means; by temptation, which is in fact subjecting the children to bribery by promising a reward undeserved; or by induction, through sincere and unpretentious normal daily behavior of the parents, allowing children to acquire their parents' ideals at the tender age when they still look-up to their parents and aspire to match their wisdom and mimic their behavior. My parents, and specifically my father, believed in the parallel development of the child's character and his mental knowledge. He read continuously and wisely, and we followed in his footsteps. He advocated intelligent learning, free thinking, training the mind to be inquisitive, and being, not acting, humble by listening to others, as there is wisdom in everyone and everything. Literature was food for the soul; poetry for the heart; art for the spirit; sciences for continued development and advancement of civilizations; and religion for better understanding what motivates the Lebanese people in order to integrate rather than dissociate. For him, no life is worth living if the mind is not free; and courage was to practice what you believe, and to voice your opinion when you are lucky to be among a perceptive audience.

Ahliah School

After a couple of years at Al Sanayeh Kindergarten, I started my schooling, like my elder sister and my younger brother, at the Ahliah School for Girls. It was the year 1955 and I was almost 7 years old. It is understandable for my sister to be at a girls' school, and in fact she continued her secondary schooling there, and obtained her BAC-1 degree before moving up to AUB. So why were we boys also there is primarily because of Madam Cortas and Miss Salma, both pillars among the best educators of Lebanon at that time. They instilled in their pupils the conviction that when armed with good education and a strong character, the world would be open for them to conquer. At that time, boys were admitted only for the elementary school, and were full fledged graduates, with a theatrical ceremony, at the age of 11 or 12. Miss Salma believed, and argued in support of the school's policy makers, that boys growing in a free mental environment would grow to be men by the age of 12, a fact that could distract the girls who are riding Ahliah's train to the exit gates of high school. That might not be utterly a wrong presumption; my father came to believe that, not only because he respected the judgment of Miss Salma, but also because he witnessed it firsthand, at the age of 20, with his first salary paying job, teaching the young ladies of Ahliah's BAC-1 class the history of Lebanon. So, we boys attended the Girls School, and at the age of 12 I was one of the four boys, among forty-four girls, to graduate from Madam Qortas' elementary school. A mathematical ratio of 10 to 1, and more so in my favor as I was top of my class, was something to thank my father for.

A Time for Reflection

Three months separated my graduation from AUB and my travel to Qatar, my first job, and the first of more than 40 years of living and working away from Lebanon. Three months compacted with memories of my schooling life, the pleasant and not so pleasant ones, a life that in retrospect was worry-free, fulfilling and satisfying, with full dependence on my father's resources, which I believed were rightfully mine by virtue of being my father's offspring. One common denominator for all those schooling years was the quest for knowledge and the continuous effort to better myself, always guided by the principles instilled in me by my parents.

This was also a period of anticipation, of wondering what lies ahead in the second phase of my life, when I have to assume full responsibility for my decisions and actions; to earn my own living and know how to spend my money; to plan for a future that is still completely non-charted; and mostly to understand my feelings towards Vicky (my future wife).

The door to the future was the job opportunity that was offered to me by Michel Malek, a friend of the family, and at that time owner of a successful contracting company in Qatar. The window to the past revealed a series of events that overlapped and in some cases were so intertwined that they remained a mystery for years to come. Rewinding the film of my life and taking a look at its events from different perspectives always had a calming effect on my soul, and gave me the chance to magnify the happy moments, to suppress depressing moments, and to invent justifications and excuses for all bad occurrences, whether they were my own deeds or those that were an act of God.

Remembering the Past

Some children are blessed with the ability to act their age, at every age in their growing up process. For them, the body and the brain mature together at the same pace, allowing them to be foolish when they are expected to; to be mischievous when mischief is a natural behavior; to be cruel when cruelty is understood even if not condoned; and mostly to be carefree to the limit of recklessness when at their age such behavior is excused. I grew up always more mature than my age demanded. Why, I do not know, and years later decided it really did not matter why. I always found socializing with my elders to be far more fulfilling than with my peers. Not that I was a parasite, clinging to un-wanting elders, but actually enjoyed extracurricular activities that were more interesting to those who were five or ten years older than me, than to those who were my age.

Monsef

Both my parents come from Monsef, a village about 45 kilometers north of Beirut, the capital of Lebanon. It is central to and the largest of six other main villages:

Berbara, Bikhaaz, Gharzooz, Shaykhan, Jadayel and Rihaneh. The seven villages are spread over five hills that run in a West-East direction, rising from sea level to almost 650 meters. These seven villages are known collectively as 'Qornet Al-Roum' being the only quarter in the Qada' of Jbeil whose population are Christian Orthodox (referred to as Roum); a small minority sandwiched between Christian Maronites and Muslim Shia's. Later on, with population growth and expansion of urbanization, few other communities came to existence over these five hills. Some became new villages, like Hesrayel, and others were extensions of the main seven. These villages were bound by intermarriages, similar religious affiliations, and overwhelming dedication to volleyball, which was the most popular sports game in Lebanon over five decades, until basketball took over in the eighties, thanks to the spreading influence of the NBA of the United States.

CSM

Monsef Sporting Club, the French abbreviation is CSM, is the first volleyball-dedicated-club among these seven villages and towns. It started as a dirt court that was constructed by the youth of Monsef, my father included. Over the years, thanks to the gigantic efforts and full dedication of Naim Naaman, it grew into a clubhouse, basketball court, stadium, den for the scouts, library and a full-fledged clinic offering free medical consultation and medicine to whomsoever needed or requested them. The tradition of community contribution, in money or human effort, claimed responsibility for the continuous additions and expansion of CSM. All members of Monsef, young or old, contributed to the welfare of the Club. They were the volleyball and basketball players; the ushers during organized games; members of the various committees that organized anything and everything, from dancing lessons for the kids to card-playing nights for the grownups, from the yearly general ball to the Club's yearly open festivity celebrating the foundation of the club, the date it was officially licensed to operate.

Scouts

One important part of the Club and the Monsef community at large was the CSM Scouts. It was established in 1970 by a handful of young enthusiasts, among them my brother Amin and Vicky's brother, George, who was its first leader. The CSM Scouts was for Monsef a very serious matter that was never shy of financial contributions, well organized, rich in activities, and embraced almost all the youth of Monsef. It was a melting pot where all social and family differences were well mixed, the rough were smoothed, the tough were controlled, and the weak had their courage and self-reliance boosted. This coherence which was instilled in the youth of Monsef, primarily within CSM Scouts, played a major role in keeping Monsef as one solid and indivisible community during the Lebanese internal war that erupted in the year 1975.

Theater

CSM Scouts is not the only promoter of community-bonding in Monsef. It is a manifestation of it as so many other facts and acts. The CSM Theater is one such factor; a cultural event that highlighted the summer-long activities of Monsef population, and reflected the "Monsefites" dedication to learning, knowledge and culture. The Theater presented one play at the end of each summer, after long and tedious series of rehearsals, arguments, and some soon-to-be-forgotten fist fights. After all it was more like a family project, where all the participants were either blood relatives or related through marriage. I remember at the opening ceremony of one of these plays, I believe it was "The Fall of Granada", my father, who was the honorary president of CSM, delivered the welcome speech and introduced the actors to the public: "My wife, my brother-in-law, my niece, my cousin, my nephew, my nephew, my nephew (my father had seven siblings) my wife's cousin, etc.., and last but not least, my son." I was that son at 8 years of age and at the only opportunity I ever had to parade on a wooden stage. The audience laughed at the joke, yet they somewhat envied the one-family village. The only person not to enjoy light humor at such an important cultural event was Uncle John, my father's cousin and long-term mayor of Natchez- Mississippi, USA. After all, he came all the way from the States, at the age of more than 70, which he would not admit to, to direct the Play of CSM Theater. So he says. But the truth is he, like most of Monsef long term immigrants to USA, could never detach himself from his roots, from his youth, and from the nostalgia for the good old days.

Monsef School

Monsef National School had a lot to do with the inherent conviction of the Monsefites in the importance of education. It was a private elementary and secondary school that survived its predecessors of public schools, and was, until the recent past, one of the very few English language schools along the coastal cities north of Beirut. Its founder, Moallem (teacher) Adeeb, whose name was synonymous with the school, is a legend unto himself. He continued the legacy of the pursuit of knowledge that Monsef was famous for by founding a school, running it, and teaching many of the subjects of learning. Miss Haifa, a co-pillar of the school, took advantage of being a cherished relative of my father and convinced him that I should attend Monsef School, in the summer of 1959, to strengthen my Arabic spelling and grammar in preparation of the "Certifact" government exams. Thirty years later, prior to our move to live in the States, Miss

Haifa was my children's Arabic teacher. She was also the teacher of Asa'd, the young Lebanese cum Kuwaiti man I met at the pool in Kuwait in the year 2007. Asa'd had spent the year of 1989 as a boarding student at Monsef School, along with many children whose parents sought a safe haven for their children during the madness that ravaged Lebanon. Moallem Adeeb's sons inherited the school as well as their father's dedication to education, and worked hard, along with a large number of employed educators, to bring it to par with modern teaching methods and standards.

Qayseeyee

It is funny how the people of Monsef refer to the past, distant or recent, as good old days. In retrospect, Uncle John's early days were not so good, at a time when Lebanon was not yet an independent state. It was out of the First World War and heading towards the Second World War; poverty prevailed; fear of an unknown future loomed over the young generation; and immigration, for those who survived, was luring them away from their community. Yet, a passing thought that transports the Monsefite immigrant from his front porch in Natchez-Mississippi to Al-Qayseeyee, the Monsef officially claimed private enclosed beach, elevates his soul, evaporates misery from his past, and brings a smile to an old and withered face.

Al-Qayseeyee is definitely the most important landmark in Monsef. It is a natural enclave on the rocky seashore of Hilweh, the coastal part of Monsef, which caters for all village-type water sports. Diving from three different heights off "Sakhret" Al-Qayseeyee, is a natural rock that reaches up to 3 meters above water level and provides three diving heights for three levels of skills. The most skillful are the teenagers who insist that they are not showoffs, yet they take twice as much to prepare for the dive as they spend on combing their hair. At the shallow end of the enclave, on a calm-sea day, you could drink fresh water from fissures in the rock slab at half a meter below sea level. There is the "Zohlayta" or the slide, an inclined rock at the mouth of the enclave that, on a day of rough seas, provides the male youth with the ultimate test of courage and few scars, mainly on their backs, to prove it. It also separates the Tomboys from the coquettes among the competing girls of Monsef for the attention of Monsef boys, and better so from the boys' friends and classmates visiting from Beirut.

Monsefites in Beirut

Almost all the families of Monsef had their primary house in Beirut, the place for work and schooling, and their secondary home in Monsef, where they spend weekends, holidays, school vacations, summertime, and retirement years. In Beirut they cluster in Ras-Beirut, the most cosmopolitan sector of the Capital. Hamra, Bliss, Commodore, and Jeanne D'arc are some of the streets surrounding or adjacent to a large number of universities, schools, hospitals, commercial enterprises, entertainment centers, fancy shops, offices, and everything that a person needs or aspires for. Ras-Beirut was a self-sufficient district where all your needs, from birth to death, are available in multi-color and many choices. The main artery of Ras-Beirut is Hamra Street, the Champs-Elysee of the Middle East. It is on Hamra Street that I grew up, in an apartment facing Strand Building, which we moved into the month Amin, my brother, was born. (Year 1951)

When my father decided to move from Ashrafieh area to Hamra Street he was labeled by his friends and family "crazy commoner". Our new address was the closest to Abu Taleb area, the wide cactus forest that extended to the seafront and was correctly named the "Hideout For Thieves"; it was also distinctly different from the posh Ashrafieh where the Elite of Beirut, the French speaking community, resided overlooking the rest of the Capital in a manner not so humble. We had moved to the heart of the middle class society of Lebanon, the population sector that earned a moderate income, sufficient for comfortable living, through the power of education. Ras-Beirut was the place where the Lebanese interfaced with a wide spectrum of Arab and Western nationals, and where exposed to and integrated with all their non-local cultures.

International College

The International College (IC) was an institution unto itself. An elementary and secondary school that was at one time an extension of AUB, thus named Prep (Preparatory) School, and hence maintaining its preferential treatment after its management separated from AUB. It boasted that its students were educated in an adult university-like environment, and as such were treated as mature people; at least this is what we IC students believed. Non-IC students, and more so their families, regarded us as spoilt brats who believe they are privileged because their parents could afford the by far highest tuition fee in Lebanon. As a former IC student, I strongly beg to differ. IC, in its English and French Sections, paid equal attention to development of character as to acquiring the best available education. High fees came with highly qualified teaching staff, excellent educational and sport facilities, and most importantly an open minded attitude. Students, within the student council body and among the class representatives, participated in setting and endorsing certain school policies, and were part of the decision making process. As a class representative, I once complained about a science teacher whose English language skills drastically failed as a conduit for transmitting information to the students. It took us few days to realize that "waaah" meant water. On a Monday morning, our class celebrated a one-hour free period due to the non-availability of our science teacher; he was declared missing-in-action by the action of the class representative.

My carefree days of IC came to an end in June 1967. My memories of the rivalry between the English Section (the true IC) and the French Section (labeled as Sex-Yonies, from the French word for section); of the happy hours with the athletic teacher who was famous for his superhuman actions, like diving from the communication tower at AUB beach and surfacing half an hour later with his cigarette still on fire and dangling from his lips; the yearly carnival; the music bands; and so many incidents and actions that at their time were not so important, but in retrospect are the substance of my youth in Lebanon.

1967 War

Through the first half of 1967, tension was building between the Arab Countries and Israel due mainly to the Palestinian issue reaching its boiling point. Most of IC students considered themselves part of the Palestinian cause, with very few actually taking part in its activities. Yet all of them decided to forget their scholastic responsibilities and focused on the intellectual discussion of the Palestinian homeland and it occupation by Israel. The best forum for such discussions was Marrouch, over a plate of "foul or hommus", at 8 am on a school day, and obviously the first math session of the day had to be sacrificed despite the voiced objections of Mr. Nader.

Lebanese are born political experts, analysts, and final authorities on every domestic, Arab, and international issue. Arguments are long and heated; group discussions are conducted with every one speaking at the same time since listening to the opinion of others is irrelevant; all issues are of equal importance and deserve the loudest possible voice; and, substance withstanding, to disagree is to be recognized as an intellectual analyst. By the time coffee cups are empty, and the peripheral listeners have dissipated, the subject matter of discussion is long forgotten, and the panel is dissolved. The forum for political deliberation and passing judgments on ultimate truths could be in a taxi (better known as Service, pronounced Serveece), at Marrouch, or at Uncle Sam. For the students of Lebanon, and more so for those pretending to be students, Uncle Sam was the place to hang out at. Over an American coffee, a burger when the financial situation permits, or better with a glass of water and a cigarette for those who believed smoking is a symbol of maturity, world politics is usually discussed when no girls are around. In their presence, the subject quickly changes to the latest movies, the music bands, the heroics of IC students, and if lucky the latest conquests in the field of the opposite sex. Those long meetings seemed important at the time; however, looking back they are what they were, an attempt by teenagers to be seen as adults when they still had long way to go.

During my days at AUB, as an Engineering student, my visits to Uncle Sam became much less frequent and hanging out was not an option when passing your exams was a serious consideration. Yet, when I managed to steal few minutes away from our frustrating study program and meet some of my "Artsy" friends (referring to School of Arts and Sciences), the atmosphere was that of the IC days all over again.

The June 1967 war started and was all over the information networks, from radio stations to newspapers, from students at the sidewalk Cafés to taxi drivers, and mainly at the barber shops and hairdresser saloons. The Group Seven, Mounah, Basem, Bassam, Samir, Rizk, Marwan and I, decided that Fayrouz nationalistic songs were more important than studying for the BAC-2 government exams. Afterall, Fayrouz is a symbol of national and Arab identities. If she sings to Al Quds (Jerusalem) and delivers in her heavenly voice the promise of return to the City-Of-Peace as Arab conquerors, so why shouldn't we believe her and bask in a glory that the Arabs so desperately needed, especially after more than 400 years of Ottoman occupation. A month later, it became common knowledge that the war was over almost as soon as it started. Glory turned to frustration, then anger, then as always, blame of American Imperialism and Western Conspiracy. Why we never admit our failures, our weaknesses, our inabilities, is beyond my comprehension. Maybe it is the best way to preserve our pride, individual and national, when too weak to protect our land, countries, and families.

Bassam's family had a summer house in Broummana town, a summer and in later years a year round haven for the Lebanese as well as other Arab nationals who sought a mountainous city/town, with perfect climate, not so high to be cold and not so far from Beirut to be away from all the action. We, the seven of us, concluded that with only one more month ahead of us for the government exams, which the Lebanese Government decided to hold on time and as scheduled, the best course of action is to retreat to Bassam's Broummana house and focus on what was best for us, that is intensive review and study in preparation for the BAC-2. Somehow, only Bassam, Mounah, Samir and I ended in Broummana, and could only last for few days before dispersing and returning to a more disciplined environment. Broummana, because of its proximity to Beirut, the dwellers of Beirut sought its active night life to vent their frustration caused by the disastrous defeat in the June War. We fell to that temptation and ended up sleeping during the daytime and consoling other Lebanese in the cafés and restaurants at night. To reverse our sleeping habits, Mounah suggested we drink lots of Turkish coffee in the evening to overcome the urge to sleep, and thus benefit from few early morning hours of studying. The attempt turned out a complete failure and a phenomenon that defied the laws of caffeine. It took new inventions in torture to wake up Mounah, and only succeeded after he managed to go through 14 hours of non-interrupted sleep. The remaining three weeks prior to BAC-2 were spent under the strict supervision of my mother, who allowed me three breaks for food and six hours of sleep in every 24 hours of utmost panic.

The retreat to Broummana was one of my early attempts at being a manager by trying to practice time management. I failed not because I lacked the skills, but because I used the wrong tools; Broummana night life is not inducive to time control. Three years later, as an engineering student at AUB, I experienced another failure at time management. Without properly assessing the scope of work and the time requirement for the courses of third year mechanical engineering, I mismanaged time allocation for studying and for entertainment. If it were not for a last minute sprint, my past thirty-six years as Mechanical Engineer could have been spent in an altogether different field than engineering. Those two experiences taught me, even before embarking on my future career, that time management is a major factor in the success or failure of any project or endeavor.

Group Seven, and all my friends at IC passed the government BAC-2 exams, thus managing to keep IC's success record intact, which never fell below 98% passing. Now, I was ready to change camps and go to AUB, which by the way is separated from IC by a narrow access road with a landmark known to all AUB/IC students as Mehyo. He was an old man who never aged; provided us with our schooldays nutrition of chocolate, Ka-ek (a kind of crispy and thin pocket of bread filled with toasted sesame seeds and thyme), and cigarettes for the smokers; assumed the role of personal message center for everyone; and most importantly extended credit facilities for buying his products. The only competitor to Mehyo was Abou Al-Miche who ran the hangout equipped with Babyfoot and Pinball machines, and was strategically located next to El Tarazi, the famous sandwich place. Those are few of the unforgettable places and non-academic people who shaped our life at IC, and separated the nerds from the cool guys. Yes, even I, whose definition in my son Majed's dictionary is "An Ancient Being", lived in an era when the word cool was correctly used.

AUB

Why did I attend AUB? That was the natural progression for any IC student. Applying to the Bechtel Engineering School at AUB was by choice and an easy decision. As an IC student who was good in math and sciences I was expected and mentally conditioned to follow in the footsteps of my predecessors, by choosing between engineering and medicine. Since as a child, as a young boy, and later when in my teens I had the inexhaustible urge to dismantle any mechanical machine or piece of equipment that exhibited the slightest malfunction, and in most cases managed to put it back together and in operation, even if not as it was intended by its manufacturer, I was destined to become a mechanical engineer.

And there I was, one of more than one hundred and eighty students, sitting in the large auditorium to be welcomed by Dean Ghosn and to be advised that not more than 50% of all first-yearers are expected to graduate, and he was proved right after four full years of tough competition, marred with bitter-sweet memories.

Summer of 1967 was a transition period between carefree life and start of serious future planning. What I chose for my studies at AUB would define my future in terms of profession, country of work, potential career, financial growth and position in society. Little did we know then that what you study at the University and what you practice in future professional and social life could be diagonally opposite, and change of course is always a possibility. So, to make the best out of this long transitional vacation, except for the period we were stuck in a military barracks in the South of Lebanon, presumably serving a month long military training, Adonis and I practically spent every waking hour either hunting birds or swimming at Qayseeyee.

Adonis and I were called in Monsef: Sarah and Mayneh. The story goes that these two ladies were inseparable and monopolized each other's friendship; they were soul mates who did everything as a pair. Adonis was not a class mate and did not even go to the same school or university that I went to; he even was part of the French system of education. From the tender age of seven, and I am only 10 days older than he is, until graduating from high-school, we spent almost every day of our vacation and holiday time together. We shared the same passion for daylong walks through the rough valleys and mountains around Monsef, mostly carrying 9mm cartridge guns that could in theory shoot and kill birds on those rare occasions when we managed a precise aim. We went to the movies together, played with other friends "Cowboys and Indians" but always on the same team, played all sorts of board-games, and conquered every rock at Qayseeyee. Lunch and dinner were always shared, at his house or ours, wherever a more appealing meal was being served.

Through my years at AUB and later during my crusade in the Middle East, Europe, and the USA in pursuit of higher achievements in the business world, Adonis and I grew apart as each went his own way, following our dreams. Yet, whenever we met, it took only few minutes to bridge a time gap of sometimes more than a year. Most importantly, it was understood and accepted beyond any doubt whatsoever that either one was available for unconditional and unquestionable support for the other, without having to ask for it. We, my family and I, were living in Paris in 1985 when Adonis called from the USA to advise that his bride of less than a month was still waiting for her visa to the Sultanate of Oman, and he had to travel back there where he was employed. He did not ask but informed us that May, his bride, would be traveling to Paris to stay with us for as long as it takes for her visa to be issued. Only after May arrived in Paris, to our door step, did we realize that we had never met her, we did not know her and funny enough it had not occurred to us that that was a bit unconventional. She was Adonis' bride, which was enough for him and us. Adonis was there for us when our eldest son, Marwan passed away; I was there for Adonis when he had to go through regular blood transfusions at the American University Hospital in Beirut for a rare disease that he eventually capitulated to at the age of 53.

The summer of 1967 came to an end leaving behind many sweet and few bitter memories that keep surfacing from the sub-conscious to the conscious mind at random without any effort on my part. Among the best things that happened were our passing BAC-2, with Samir ranking the 2nd in Lebanon, and my first realization that I was attracted to Vicky, and not necessarily in that order.

Samir and I can claim that we were classmates and close friends from pre-school and forever. All through elementary school at Ahliah I ranked first and he ranked second in class. I was the outspoken and socially fearless one (by the way I never changed in that respect) and he was the quiet and serene person. I started high school at IC and he went somewhere else, to join me two years later and continue together, through IC and the mechanical engineering department of the Bechtel School of Engineering at AUB. Till this date I do not know why he spent those two years, 1st and 2nd secondary, at a different school. The only thing I know is that he must have gone through a brain restructuring, which manifested itself in him getting ahead of me, school-ranking wise, at IC and AUB. Our friendship must really be strong to survive this switch without either of us caring about it.

Vicky is another issue altogether. Her father comes from Monsef and her mother from Jadayel, the adjacent village. She was a tomboy, whom I had not noticed until the summer of 1967, although she was a friend of my brother Amin, and her father was frequently my partner at a card game we called "Leekha", better known now-a-days as Hearts. As far as I am concerned Vicky came to existence as a model parading in her purple hot pants at a charity event held by CSM, the sporting club of Monsef. That summer, even though I preferred to move around Lebanon driving my father's car, since I had just turned eighteen and got my driving license, still I joined the Monsefites on the bus rides, accompanying the CSM volleyball team in their never-ending loosing championships, only because Vicky would be on those bus trips. I suspect she knew I had ulterior motives, other than watching a volleyball game, yet she pretended not to notice. And I was hooked long before I knew it.

Hiding my emotions and suppressing them from facial expressions was a weakness that I could never control. I never had a poker face and was never a poker player either. This lack of control remained a deficiency in my management career. My disgust, dissatisfaction, and anger, specifically when business is concerned, surfaced so fast leaving me no time to put a lid on them. I could always tolerate ignorance in others, and often did something about it, but never would I accept stupidity which stemmed out of carelessness.

My life at AUB started in October of 1967. For four years, full years including summers, most of my friends and I had one and only one goal: Stay in the race, push our way through, and reach the finish line. We were the ones destined to enroll in the Engineering School or the Medical School. The few who went to the School of Arts and Sciences, the Artsies, to major in Business Administration, Economics, and Political Sciences had other interests all together. It is true that they wanted to eventually graduate with a degree, however time was not of essence. Those were the friends who, through sit-ins, strikes, demonstrations, cutting classes, and occasional drug trips, claimed to be the conscience of the Arab Nation (Al Oummah Al Arabiah), the guardians of Arab nationalism, and the determined-to-be liberators of Palestine. Few of my other Artsie friends went through the same behavioral techniques as the Arab nationalists, but their cause was completely different. They believed that the Lebanese are Phoenicians and not Arabs, and thus the Palestinian cause, Arab nationalism, and all those Lebanese who believed otherwise were dedicated to destroying Lebanon, which is a piece of heaven on earth. The conflict between these two major forces kept AUB Upper Campus alive; AUB milk-bar a stage for political debate; Uncle Sam a forum for intellectual prophesies; and most importantly, kept these students as students for far longer than the curriculum requires. We, the engineering students, were entrenched at the Lower Campus with one enemy to fight, the system that guarantees a 50-50 chance for failure and a resulting re-location to the Upper Campus.

Not all Engineering students were bookworms. Some, like my friend Samir, had a natural rapport with engineering material; others were satisfied with just coping rather than joining the rat-race; and few, including myself, tried to maintain a balance between enjoying life and seeking technical knowledge and good grades. However, all of us students had one common factor: Fear of being beaten by the rigid system of the Engineering School, and despairing before failing or graduating.

Engineering School

In the late sixties, the Engineering School followed an educational methodology not much different from high schools. We had no elective courses; we could not choose our course-load per trimester; we could not drop courses or rearrange their sequence; and the worst thing is that we were allowed to fail two minor subjects, one time only during the 4-year program, provided we obtained the passing total average for that trimester, and had to wait a full year to re-take that course since it was given only during the same trimester every year.

First-year engineering was a hotpot with all students clustered together, fighting to melt-in rather than float and spill over. During the third trimester, those who survived the first two trimesters, ended up at a large camp area in Mazboud, a village up in the mountains and south of Beirut. We actually lived in tents; answered nature's call at ablution blocks; and ate in a mess-hall tent a weekly menu that was more of a test for survival than a means for survival. The Mazboud camp was the location for the obligatory course of land surveying, a subject that the mechanical engineers-to-be regarded as a waist of time and mismanagement of our potential. We were wrong, which I realized few years later when I was assistant project manager on a pipeline project in Qatar. I had to live at a project camp, among a grumbling bunch of the toughest professional welders, and had to manage them, their daily personal problems more than their work, to keep them on the job and to keep the project going. At that later date, it dawned on me what Mazboud camp was all about; that it was a lesson and a shy training in management of a large community of diversified professionals, where nagging could drive the less experienced into resigning from the job and from the engineering profession all together.

My days at AUB came to an end in June 1971. Life at the Engineering School, from the start of the second year to that end date, was a monotonous regime we had to endure to be rewarded with a Bachelor of Science Degree in Engineering. My memories of those days are tainted with one overriding nightmare, the fear of failure. Not that the Engineering profession was a passion or my only chosen mission in life and I desperately wanted to be an engineer, but failure itself, and most important its impact on how others perceive of me, was what I dreaded most. There must be a psychological explanation to why I could never tolerate criticism;

why I believe my comprehension of any issue that I consider myself knowledgeable in should not be disputed; that my logical analysis of any subject is flawless; and why I needed to be appreciated for my contributions, where a pat on the back meant far more to me that monetary reward. *I have always practiced self-criticism, self-evaluation and self-assessment in my belief that these lead to self-improvement, and always came to the same conclusion that fear of failure was not caused by my insecurity but by the dire need to achieve and to be recognized for my achievement. The importance of expansion of knowledge, the urge to achieve and the significance of reward are basic principles that I adhered to throughout my management practice. I even encouraged others to seek and accept criticism; however, that was something that I preached but failed to practice.*

I remember during the first trimester of my third year I decided to quit engineering altogether and go for some other profession. Memory of that lousy chapter of my life is cloudy, thanks to the self preservation mechanisms of the human brain. My grades were bad, my interest was worse, and my will to persevere was almost gone. My family supported me all the way in whatever decision I would have taken, however my father's advice was not to succumb to the fear of failure. "The greater success always comes after an interim failure", is what he said and what I witnessed first hand in the evolvement of my future career. I reversed my attitude towards engineering and finished the 4-year program with a reasonably good overall average grade. Of course, all that fear dissipated during the last year when graduation had become an established fact and only a matter of time.

Graduation Ceremonies

All my friends made it to the end as well, and we all looked forward towards the graduation ceremony and the stage exhibition in flowing gowns. When this did not happen, I believe I was the least disappointed among them. After all, fate had already conspired against me at each milestone in my schooling life in order to deprive me of the pleasure of on-stage parading. During my last year as an elementary student at Ahliah, Madam Cortas decided to restrict the graduation ceremony to the high school students due to severe political tension that was broiling in Lebanon. Few years later, in 1965, the Lebanese Government decided that as of the following year all Lebanese nationals must pass the BAC-2 exams to be admitted to Lebanese universities. So, in 1966 IC made the historic cost saving decision to abolish the concept of graduation for those who pass BAC-1 exams. Obviously, I was in my 6th high school year in 1966 and thus, even-though I passed BAC-1 exams, I succumbed to that ridiculous decision and left my first manually tailored dark suit hanging in an equally dark place, unused. The defeat in the year 1967 was so overwhelming that IC, the administration and the students forgot all about graduation ceremonies. In fact, ceremonies of any kind would have been improper. The last straw was the reaction of AUB to the student strikes in 1971 which, having been given enough advance warning, saved me from hanging a new dark suit next to the 1967 one. I think my worry at being the cause of a Third World War if I take another shot at graduation ceremonies, prevented me from pursuing post-graduate studies. I hope my children and their children benefit from my sacrifice and live a life free from major or minor wars.

Activities at AUB

During my first few months in Qatar, whenever the weight of being on my own and not yet equipped to clearly see my way into the future got too heavy to handle, I let my imagination fly back in time to the near past to focus on the pleasant moments that were scattered over my days at AUB. The Engineering ball, a happy dancing-intensive occasion loaded with loud noise mistaken for music, initiated every school year. We over-enjoyed it since it was very well known that that was the first and last carefree time the engineering students were allowed to engage in for the remaining 364 and quarter days of the year. Some of us, me included, managed to take short and not so frequent trips to fun-time world in between dull steel and concrete classes. Escaping every now and then to enjoy an hour of Table Tennis was important for me. It was more than a game or a sport or even an escape from the dry engineering courses. Table Tennis was the second most important sports activity at CSM, after Volleyball, and it did not require the physical characteristics of the latter, tallness from foot to head, which I lacked. However, it required sharp reflexes, a quick mind, and most important perfect coordination between body movement and the mind. At the tender age of 10, I discovered that I had a built-in well developed reflex mechanism that allowed me to grab a suddenly falling object before hitting the ground; respond quickly and correctly to out-of-the-blue remarks, especially if they were sarcastic or induced a sarcastic reply; and to talk myself out of and embarrassing or tough situation where quick reflexes saved me from consequences that, in the absence of physical strength, could not be otherwise avoided. *Table Tennis was instrumental to maintaining and improving my sharp reflexes, which through my career days proved to be a very important management tool when dealing with people or ideas alike.*

Swimming at the AUB beach was another escape from engineering routine. We raided AUB beach as IC students, in spite of its highly polluted water and its abundant supply of stinging Jellyfish, only because of IC's relationship to AUB. Our pathetic attempts at pretending to be more mature than our ages allowed were tolerated by the female freshman and junior students.

Another beginning of the school year interesting activity is the Welcome Party for the Engineering First Yearers. Of course it is interesting for us when we are welcoming and not when we were first welcomed. During my third and fourth years I invited Vicky to be my partner, or what is now called my date, in order to give her the opportunity to compare me as a seasoned engineer with the new tender recruits. Come to think of it, we have been married for 40 years and I still don't know how she had rated my maturity as a third yearer compared to the first yearers. I suspect she could not find much difference, although she made me feel otherwise; maybe she also subconsciously had stronger feelings for me, more than the occasion required.

Training in England

The full four-year engineering program includes a ten-week work requirement with a well known international establishment during the third trimester of the third scholastic year. We all looked forward to this mission since it was governed by international exchange programs among American Universities, allowing students to work abroad; get credits; get a feel of what the engineering profession is all about; earn an income, however meager it was; interact with other nationalities of other exchange students; and most of all enjoy the satisfaction of having reached so far that graduation was assured. My turn came in the summer of 1970, and my work place was Hayward Tyler Pump Division in Luton, England. Landing that temporary training job was not through AUB exchange program but through the help of my father and his friend, who was the head of the textile industry in Manchester and the British Employers delegate to the International Labor Organization in Geneva. *Important Lessons in business: Make good connections; offer and give help and assistance when you can without asking for compensation; be and not act sincere about helping others; and be thankful when people return your favors. As long as there is no material or money involved in asking and receiving favors, and as long as you stay noticeably ahead in the exchange of favors, most of the people who possess sound human values would be glad to reciprocate.*

I took the train from London to Luton and then a taxi to the guesthouse which was prearranged by Hayward Tyler Co. It was a nice townhouse, managed and serviced by the owners, an elderly couple who provided their paying guests with bed, breakfast, dinner, and free home laundry service; underwear and socks were excluded. Each room had two beds but one guest per room, unless there was sudden need to accommodate a one-night passing by guest and no vacancy was available. As English manners dictated, permission from the more permanent guest was first taken. And so it was that a UK guest happened to share my room for one night, spent more than two hours telling me stories, and then went to sleep before I could decide whether he was Irish, Scottish, Welsh, or just spoke cockney English, in order to try and understand what he was telling me. Unfortunately, he was gone the following morning before I woke up, and left me with a mystery never to be solved. There was another mystery that I luckily solved before it drained my well managed tight financial budget. Laundromats handled my underwear but, as my mother had instructed, socks are too delicate and had to be hand-washed. So, on a Friday afternoon, after a fortnight in Luton and fourteen pairs of socks marinating in a plastic bag under my bed, I came up with a bright idea that I was sure would achieve good cleaning results without wasting precious weekend time. I filled the wash basin with lukewarm water, you see hot water destroys the elastic band, emptied a generous portion of washing detergent and worked it into a beautiful foam with the socks totally submerged. I went to London immediately after, spent a well deserved weekend, came back late at night on Sunday and walked directly to the washbasin. Apparently the rubber plug was not as tight as I had assumed it would be; the water was gone; the socks were so crispy that they would crumble into pieces with the gentlest touch; but, they were clean. The following morning, I had to replenish my two weeks stock and decided to adopt the alternative and less costly solution of using Laundromat.

My short stay in Luton was my first real exposure to Western cultures in the West and not as modified to suit the Lebanese traditions. Two things remain to-date clear enough for me to believe that they must have influenced my behavior, social and professional. I received a phone call from the English gentleman who was instrumental in securing my training program at Hayward Tyler, inviting me to dinner on the following Friday at a posh restaurant on the outskirts of Luton. He was on his way to London from Manchester, and was making a significant detour to meet me. I was overwhelmed by my own feeling of self-importance, when in truth his invitation was a reflection of his respect for my father, and decided to measure up to my reputation as a soon-to-be Mechanical Engineer from AUB. At the table, my host offered to order for me in such a gentle and delicate way which the English noblemen are famous for, and I graciously accepted. Imagine my shock when after an elaborate conversation between my host and the head waiter, I am served with a glass of perfectly aromatic wine and a plate of melon. After all the anticipation and the exquisite service, not to mention my all day fasting to be ready for the big one, I was to be fed fruits, and only one kind of it. This time, my inability to hide my emotions was to my advantage. The expression on my face was clearly read by my host, and before I had a chance to make a fool of myself, he explained, again in his highly bred manner, that the proper English cuisine started with smoked ham and melon, to be followed by the main dish and the rest of the meal. *The message was delivered, without a word being said, that the world does not run by my rules or per my expectations and that it pays off to be patient rather than jump to conclusions, especially if having a poker-face is not in my character.*

Hayward Tyler Pump Division had been in business for many years when I moved to Luton as a temporary resident for few weeks. I am not a person who could be intimidated or overwhelmed by people, facts, incidents or encounters. This is how I was raised by my family and shaped by my teachers, with a crystal clear advice to not confuse these positive attributes with the obligation to respect my elders; to recognize the achievement of others; and to bring out the best in people as each person has something good that can be properly cultivated. Apparently the English technical manager, under whose wings I was to develop my first engineering skills, had the same approach towards strangers, even though we came from different cultures and opposite directions; the volatile East and the cool West. I respected him for what he was, my boss with far more knowledge and experience, and he reciprocated, recognizing me as a soon-to-be engineer who should be trained to qualify as a real engineer; and that was the basis for our instantly established friendship. *It was at Hayward Tyler where I realized the significance of Trainer-Trainee relationship and the importance of "Transfer of Knowledge" in order to contribute to the development of the human element and to the community in which we live.*

The momentum of self-assurance that I acquired in Luton during that summer carried me smoothly through my fourth and last year at the Engineering School. It is funny how my attitude differed from that of a year before during the first trimester of the third year, when I had been ready to call it quits, especially that I had spent the preceding month vacationing in Paris and London with Samir and Bassam. The vacation itself was compacted with all kinds of pleasant events that have enriched my portfolio of memories, and somehow find a way out whenever we reminisce about the carefree days of the past. That was my second trip abroad and the first as an adult and on my own, but each was for a completely different reason.

Trip to Paris

I fell in love with the Paris that was portrayed with affection by the professor who gave us a course on the history of architecture, during our second year of engineering. Bassam was also inspired by the beauty of a dream-city, and Samir …, well he just went along. Samir has been a loyal friend all my life, to the point that we have developed some form of telepathy, the proof of which is the email I received from him, after a silence of years, at the exact moment of writing this paragraph. We landed at Orly airport on a beautiful September afternoon, and by the time we reached Place De La Concorde, I was enchanted. Bassam and Samir stayed at a hostel that provided free accommodation for Lebanese students during summertime. I stayed with an aunt of mine whose husband had recently been relocated from Lebanon to be in charge of Ford Motors Company in France. The only difference between where I stayed and where they stayed was the free breakfast I got before leaving the house, at never later than 9 AM. The three of us met every morning on Avenue Paul Doumer, and walked the streets of Paris with our necks tilted upwards, swinging from left to right, admiring every arch, column, and carved structures, and often oblivious to the undesirable soft objects that we frequently stepped on. How we got the energy to walk around for almost fifteen hours daily, surviving on junk food and quick sandwiches, since we were on student budget, is a further proof of how completely enchanted we were by everything around us. The last night in Paris, we decided to treat ourselves to a well-deserved dinner after two weeks of self-imposed starvation. The decision was made on an impulse upon coming across a posh restaurant, with a doorman as rigid as the guards of Buckingham Palace, and a red carpet so long that the small passing-by French cars had to drive around it. If our tired looks and very casual clothing betrayed us, the restaurant receptionist managed to hide his expected reaction well. However, the maitre-de failed to control the twitching of his eyes when halfway through our meal he glanced towards us and found me struggling with all three dishes that we had ordered for the three of us. The maitre-de was not aware that I was the one claiming to be fluent in French and thus was assigned the duty of ordering food for all of us. I was, and still am, very good at French; but, I have never been able to tell the kind or type of food from its sophisticated dish name, in any language and more so in French. One dish of the only five on the menu had two French words in its paragraph-long name which I recognized. I translated "ris de veau" as veal and rice, and so ordered three plates with my friends' consent. Unfortunately, rice in French is "riz" and not "ris", and when ris is combined with veau it becomes "sweetbread", with no trace of meat or rice. So, for our last and only expensive meal in Paris, I ended up eating three plates of a very un-tasty substance, costing me a significant amount of my allowance, and Samir and Bassam were satisfied with hotdogs sandwiches, bought at a close by food stand, and a good laugh at my expense.

Trip to Jerusalem

My first trip abroad was to Jerusalem, at the age of twelve, when Jerusalem was still a divided city with the East part under the control of Jordan. My aunt Nelly, now that my age has already caught up with hers I just call her Nelly, was to be married in the "Church of Resurrection" in Jerusalem. Nick, her husband-to-be, is a devout Russian Orthodox who loved Nelly so much that he wanted their life together to be sealed at the holiest place of the Orthodox Church. Since Nick was at that time the manager of Ford Motor Company in Lebanon, he decided to travel to Jerusalem in style, accompanied by few selected family members of Nelly, in two Ford Galaxy cars. In addition to being assigned the important role of ring bearer, I was a friend of Nick and Nelly's brother Moufeed, age difference notwithstanding. It took forever to reach Jerusalem from Beirut, despite the few stops on the way including lunch at the Dead Sea, where people floated in a sitting position while reading the newspaper, as if suspended from above. My memory of that three-day trip is hazy except for the all-night session we spent crammed in one hotel room, with Moufeed and me telling jokes till it was time the following morning to get dressed and get on with the wedding ceremony, and then head back to Beirut. Nelly is the same Aunt with whom I stayed in Paris on my second trip abroad, but by then she had two very young daughters who thought my French was superb, and maybe were partially to blame for the three plates of "Ris De Veau" that I had to eat.

CHAPTER II: 1971 to 1976

Qatar

The journey to Qatar was the first of very many trips to come during the second phase of my life, and was the most significant as it initiated me into the global human workforce. Some time ago, and consistent with my nature to better understand facts and issues when they are compartmentalized, I developed a packaging system for my life. I classified it under three main categories: Business, Social, and Emotional. Business in divided into three phases: Pre-employment, employment, and self-employment. The first day of the second phase of this category started when the MEA plane landed in Doha – Qatar, and coincided with the first day of the Islam Holy month of Ramadan. The Social category is subdivided into two phases only: Bachelor life and married life. Vicky exists in both of these phases, and she made the transition from one to the other so smoothly that it took me a while to realize that they were two phases and deserved to have a main category for themselves. The Emotional category is where I store all my emotions, and then they freely sort themselves, with frequent shifting between its two phases: Before my son Marwan's untimely death and after his physical absence from our life. However hard I try, I could never instill enough discipline in my emotions to get them to stick to one phase, and thanks to their defiance, Marwan remains alive in my mind and heart even though he immigrated to the other World.

Fadi, Kamal, and I disembarked around noon time and were immediately hit with a hot wave of sand, known to all those who preceded us to the Gulf region as Toz. Fadi and Kamal were both my classmates at the Engineering School, but they had graduated as Civil engineers, and were to be my workmates at Orient Contracting and Trading for more than three years. Fadi and Kamal were recruited through the normal procedure of applying in response to an employment advertisement; they were interviewed and hired. For me, the path was different and it started with Michel Malek, the major owner and partner, who happened to be a close friend of the family and frequently addressed by me as "Am-moo" or Uncle Michel, offered to initiate me in the mechanical engineering profession under his patronage. An added value to my employment at Orient was that Salah and Khaled, the senior engineers and minor partners, were good acquaintances for several years prior to boarding the MEA flight bound for Qatar. Salah was the eldest brother of Mounah, and a member of a family that treated me as a son. Umm Salah, the mother of the family, never forgot to invite me to lunch when she cooked stuffed cabbages. Suhail, the youngest brother, respected me enough to accept my tutelage during his wild days as a teenager, when studying was the last thing on his mind. Suhail ended up marrying my first cousin, and Munah was my best-man and the friend-indeed when in need.

Khaled's family was a close neighbor to my Uncle's family in Ras Beirut. We knew each other by association, through my cousins and through Salah's family, and shared a common hazy memory of my graduation after-party that was held at "Nabeh-Al-Teffaha" restaurant, where half the attendees were self-invited, turning the party into an open-house for anyone who played a musical instrument, liked to sing, or enjoyed dancing, with one common factor: willingness and ability to consume large quantities of Arak, the Lebanese tougher-than-nails alcoholic drink. Khaled belonged to the group of musical instrument players, and he played the

"Der-ba-kee", an Oriental type of drum that is beaten by the fingers and the palms of both hands rather than by a stick.

After a welcome speech by Salah, and going through the necessary arrangements for new arrivals, like meeting the accounts and administration officers, we retired to our chosen living quarters to unpack and rest for a while. Fadi and Kamal decided to share a two-bedroom unit while I preferred an adjacent single bedroom studio for me, with their large sitting room serving as the reception and dining area for the three of us. By 4 o'clock in the afternoon, we were all excited enough to gladly accept Khaled's offer of a tour of Doha city. I had had a late breakfast on the plane, a meal that MEA airlines was famous for, and just grabbed an apple to munch on in the car. Khaled was excited about his new role as a tour-guide in the city, where he had already spent four years of hard work, while we just looked through the car windows wondering how we were going to survive in an environment of pure sand, pavement, and single-story plain houses. Trees and greenery were as rare as the likelihood of a policeman happening to be cruising on his official motorbike at that unruly hour of the first Monday of October 1971, just as I was biting into my apple, and pulling us to the side to give me a long and monotonous lecture on why I should not be eating in public during the fasting hours of Ramadan. I was raised in Lebanon, where Muslims, almost half the population, lived in social harmony with Christians, and where religious practices by either religion or their various sects were not imposed as overall social rules. *First lesson for me in Qatar: always learn about the cultural, social, and religious practices of any community before you move into and interface with.* Otherwise, you end up disposing of a good apple when you badly need it.

The first ten days in Qatar were fully consumed by formalities with various government agencies to legalize our residency, work permit and driving permit. The first two procedures required medical tests, blood test, x-rays, eyesight check, injections and immunization, which we vehemently resented as, to us at least, implied that either we had come from a country infected with diseases, or that the Qatari government wanted to make sure that we do not fall ill or die in Qatar and have the officials worry about shipping our bodies back to Lebanon. I now believe that those procedures were normal, as in later years they were applied to me in all the various Arab countries I resided and worked in, and that our over-reaction in Qatar was the result of homesickness. The driving permit was a totally different and somewhat humorous experience. A holder of a Lebanese permit was only required to pass an oral test which, we found out at the last minute, was strictly the correct identification of three traffic related signs out of a list of about fifty. I don't remember if I volunteered to be the first to approach the man sitting behind separation bars in a small booth at the traffic department, or if Fadi and Kamal volunteered me, but there I was looking at a chart with a finger, not belonging to me, pointed at a triangular shape with what looked like a distorted picture of plane in its middle, and two dark lines drawn underneath the plane. I immediately realized that I had no idea what that sign meant, and that I was going through the same experience of few years earlier when I had applied for a Lebanese driving permit. The first thing that came to my mind, and simultaneously escaped my lips, that the sign indicated a parking area for planes. The man inside, in all seriousness, responded by a single word: next. I don't know what Fadi and Kamal's responses to the same sign were, but the result was that the three of us failed and were asked to go study the list of signs and to come back one week later. I took the liberty of asking him what the sign stood for and he, maintaining his aura of importance, said: A plane flying at low altitude. I should have stopped there and backed off, but my natural instinct for sarcasm took over, and asked that if we encounter such a sign while driving a car should we stop and duck under it or drive the car into a ditch to avoid being run over by the plane. Luckily, I ducked in time to avoid a collision with the pen that flew towards my face, and managed to catch up to my fast-running colleagues before our official friend let himself out. *Lesson number two in Qatar: humor is often wasted on government employees, especially if they take their job with more seriousness than is warranted.*

Salah assumed the responsibility for indoctrinating Kamal in the civil building construction; Khaled was to be the mentor of Fadi for road construction, and I for the plant department. I remember Michel Malek telling me over dinner at his house, after a couple of weeks into my three-year duty in Qatar, that heavy construction equipment was the most significant element in infrastructure contracts, and a mechanical engineer had the advantage of better managing this equipment. And so, I was assigned for full six months to an office close by the garage area, full of catalogues, operation manuals, and maintenance instructions for all types of heavy construction equipment and vehicles, and had only one responsibility: Get to know each machine, each system, and virtually every nut and bolt. Gradually, I could tell the difference between a Caterpillar D-8 and D-9, a shovel and a back-hoe, a rough-terrain and a wheel tractor, and a fixed-jib and a hydraulic-jib crane. But most of all, I was able *to participate intelligently in the exercise of assigning what piece of equipment to do what job at what construction site, for a most efficient and cost effective utilization* of the relatively large fleet owned by Orient Contracting and Trading.

Plant Department

In recognition of my meeting the educational goals set for me within the allotted six-month period, I was assigned by Khaled (many years later I came to know that it was instigated by Michel Malek) as an assistant to the chief mechanic of the vehicles garage. I was furious and wanted to pack and go back home, but Khaled, who had become more of a friend than a boss, calmed me down and said: Where do you find a company that is willing to pay you an engineer's salary while asking you to work as a semi-skilled mechanic? I remembered then the words of Dr. R. Ghosn when he launched us as fresh engineering graduates, and decided to climb the ladder of my profession one step at a time, starting from ground level. In a short period of time, oil and grease lost their foul smell; rubbing elbows with laborers, mechanic helpers, and full fledged mechanics was as natural as accepting their comments and suggestions; and I became a member of a buzzing team that worked day and night to meet the needs of the construction projects. By the time that I could compete with the head mechanic in diagnosing the problems of a coughing truck engine, I had become an active member of the more sophisticated garage for heavy equipment. On the technical side alone, the magnitude of problems, the complexity of the solutions, the number of operating systems, and the significance of time element, were overwhelming. The more complex issue was the conflicting relationship I had with the technical staff of the heavy equipment division. On one hand I was an apprentice and their trainee, and on the other I was their boss. My reward for my performance and achievement at the end of my first year with Orient, in addition to a hefty bonus that did not last long, I was assigned as the head of the plant department, with all equipment and vehicles, garages and their staff, maintenance crews, stores, and local procurement reporting to me. Khaled directly and Michel Malek indirectly kept a watchful eye, more for my benefit than for the benefit of the work. I was so proud to be admitted to the management club after only one year of joining the business world, and in my naivety I thought a boss is simply a person in charge with a staff taking instructions from him, and most importantly being called: Boss.

Exercising Authority

I had been a boss for less than a month when I faced the first defiance to my authority. George, the Lebanese heavy duty equipment chief mechanic, who was also physically heavy, was in my office being sternly reprimanded by me for being late in repairing a bulldozer that was badly needed on one of our sites. *He had given the store keeper a wrong part number for a defective oil seal, a small error that in the construction industry could be the cause of costly delays, wasted manpower time, and maybe contractual difficulties.* I must have overwhelmed him with my exaggerated reaction that he shifted from a defensive attitude to the offensive, and asked me to step out of the office to teach me a lesson I never wanted to forget. He stepped out, I stayed in, and as he did not expect me to follow, he went away fuming. After regaining my composure and recovering from the shock of being challenged as a boss, I went straight to Michel Malek and asked him to terminate the services of George and send him back to Lebanon on the first available flight; otherwise, he would have my resignation on his desk in an hour. His immediate response was: As you wish, you are the boss; and that diffused me instantly. He asked me to stay for a cup of coffee and to calm down, and then in his fatherly manner lead me to a discussion about the plant department, how efficient and quick was our response to the needs of the construction sites, and somehow we got to the performance of the staff. By that time the doors of reason in my brain that were shut by anger had gradually opened, and Michel Malek had noted that before I realized it. Suddenly his comment, after taking a quick and discrete look at his watch, was: *Take it easy for the rest of the day; think about the effect that George's immediate departure would have on your work; assess the impact of losing your best mechanic in the Company; see if you can salvage the damage in your relationship with your employee without losing your authority; and in the morning take the action that you deem best,* with my full support upfront.

I had not realized that a boss had to consider so many factors before making a simple decision regarding the termination of an employee. The hours of that night went by with me chain-thinking of ways to resolve my first management problem, when neither the Engineering School nor my limited work experience had prepared me for dealing with an issue that involved a personal factor. At 10 AM the following morning, I sent for George to come to my office and still had no idea what I wanted to say or the action I should take. I had decided to just react to the encounter and let my reflexes kick in for the decision. The instant he walked in he acknowledged his improper behavior of the day before, which he attributed it to work stress and personal pressure he was under, and requested in a very genuine manner my acceptance of his apologies. Without addressing his request for apology I asked him to sit down and explained to him that our business relationship is similar to that of a father and his son; there were many things that I knew better than he did and others he knew which I did not; my role was to guide him and help him and in return he had to accept my authority and respect it; in our daily dealings, if I over-react under the pressure of work demands he should realize it is not personal, and I, in return, should be more sensitive to the feelings of my employees. Only then did I realize that his anger was mostly directed at himself for his inability to read or write in English, and having to hide this and ask his helper to read the part numbers while he scribbled the wrong letters and numbers. And so, I gave George two options to choose from: either pack and go back to Lebanon or accept to spend one hour every evening with me learning English, an offer that he could not refuse. Six months later, George was promoted to head of the maintenance section; he had developed forms and procedures, all written in his simple yet understandable English; and, was recognized as the nerve center of all projects that relied on heavy machinery and equipment. To this date I do not know if Michel Malek had talked to George after our session at his house, and I do not think it is important to know, because I handled my first management crises to my satisfaction and to the expectations of my mentor.

Project Management

The incident with George set me on the path that I decided to follow in my career, the road leading to professional management. With time, I came to believe *that management is an art that you are born with a flair for; a skill that had to be continuously polished; a science that needs a lot of studying; a complex issue as diversified as life itself; and a life built on experiences, with each adding a new lesson to be learnt and stored in an active file. In short, management is as complicated as its main and primary elements, the human being. No two persons are alike; no two people react similarly to the same stimulus; no individual reacts in the same manner to different acts; and it is rare that the same person reacts in the same way to the same incident at two different times. I knew that I possessed the basics for a career in management, but had a long way to go to develop it. The ease of my interaction with others, whether friends, relatives or complete strangers, and regardless of age, education, profession, gender or nationality, was so natural that it quickly dissolved any behavioral barriers which hinder smooth communication. Courage to be outspoken and to voice my opinion was something that I never lacked. Some might call it being argumentative, opinionated, or even annoying; however, I always managed to articulate and to defend my point of view. From my early childhood, I was raised by a family that did not believe in classifying people by race, gender, the so called social status, or any other factor that puts an individual on a higher pedestal than others. People do differ by factors beyond their control such as in-born characteristics, intelligence, artistic temperament, patience and rich vocal cords; by genetically inherited physical features that result in blue eyes, dark hair, or muscular structure; or by naturally occurring alternatives resulting in being one of the two genders, or with dark or fair skin-colors. These factors should not be conducive to preferential treatment. What distinguishes individuals is the will to learn; personal achievement; aspiration for knowledge; perseverance when faced with difficulties and obstacles; loyalty and compassion to one's ideals; and most of all the courage to stand up for one's ideas and beliefs. When acknowledging the inevitability of these differences and the reverence of honest achievement, interrelationships are reduced to acceptance of the former and respect for the latter, and individuals are regarded as humans and not holy beings.*

Communication

"To manage you need to maintain sustainable communication with others"; I believe Khaled quoted this statement, as he often quoted other enlightening statements, to encourage me to immerse myself in heavy reading in search of such insight to practical knowledge. Diversification and intelligent selection of reading material and subjects are essential for developing communication skills.

Psychology and sociology improve the ability to understand how to talk to others; art, literature, and music provide an excellent entry to start a conversation with professionally or socially successful people; raw technical and basic scientific information instantly attract those who are involved in the industrial field; staying current with politics is a prerequisite for any conversation longer than ten minutes in the Middle East; History sharpens the ability to intelligently analyze current events; geography compensates for lack of travel; and general information, obtained from reading international magazines, newspapers, and any reading material that happens to be within your reach, fills in the silent gaps with non-focused or shy audience or poor conversationalists.

Reading

Reading for me is a habit acquired at home and institutionalized during my school years at IC. My father was a legendary reader from his early youth; at the age of ten he would be caught at the early morning summer hours with a fishing rod in one hand and a book in the other, when his catch was essential for feeding his siblings. At the age of fifteen he was always a winner in any poem-reciting challenge; during his high-school and university years, for lack of money, he tutored his colleagues in order to read their books; and in later years he found in reading a refuge from creeping frustration while he waited for my mother to get ready for their evening engagements. My mother never carried a watch and adamantly refused to accept time as a factor in her daily life. She did what she had to do when she was ready for doing it. If her action turned out to be in a timely manner, that would be a coincidence that meant nothing to her. Books around our home, even in the smallest room where we responded to nature's call, were so readily available it encouraged a subconscious reach for them to fill in idle time.

At International College, I joined the Library Club as a member and later as Chief of Stacks, to mainly develop skills in selection of reading material and reading habits. Diversification and sequencing of subjects are important to maintain a high rate of absorption of knowledge. I got into the habit of parallel reading: Switching between two books on completely different subjects, like a science fiction novel and a book on management in the construction industry. This technique reactivates auto-focusing when my mind starts to drift away and the words are reflected by my eyes with so few passing through that their meaning fades away. However, if this system is abused, or if switching time is induced by external factors, like a running TV show, rather than by being controlled by mind-wandering caused by a monotonous subject matter, the reader ends up drifting away from the mainstream of either book, and resorts to watching the TV.

Interview with Bechtel

My stay in Qatar lasted three years and came to an end by mere coincidence. I was visiting Moufeed, Nelly's brother and a first cousin of my mother, who was at that time the UK area manager of TMA, the Lebanese cargo airline. Moufeed is a few years older than me and a friend from the days when he was in his late teens. During a dinner party held in my honor by Moufeed and his wife Meesha, I met an elderly American engineer who was a senior employee of Bechtel International. We got into a discussion about contracting and engineering in the oil and gas industries. He was a veteran in that field and was responsible for building up three engineering offices for Bechtel, in London, Paris and Beirut, that would spearhead the newly established Oil and Gas Division. I had just completed the construction of the natural gas pipeline, running between Dokhan and Umm Said in Qatar. Beirut office was to be the design center for the Middle East and Africa projects, and he was recruiting local engineers to supplement the professionals relocated from other Bechtel offices worldwide. I was given a name to meet in Bechtel-Beirut, on my way back to Qatar, to be interviewed for possible employment. Although my employment with Orient was gratifying, I was ready to entertain a job in Lebanon, especially now that Vicky and I were nearing the point of tying the knot and Lebanon offered far more than Qatar in terms of social life and entertainment.

I landed in Beirut on a Friday night and was flying to Qatar on the following Sunday morning, leaving only one day for all social obligations and quality romantic time, as well as the squeezed-in interview. On Saturday at 9:50 AM, ten minutes before my scheduled interview, I arrived at Bechtel's offices near Beirut Airport. The doors were open but, other than the cleaning lady who was going about her business, there was no one else. I have always maintained two rigid rules regarding appointments: The first is to arrive 10 minutes early, and the second is to leave 15 minutes after the scheduled time if the other party failed to materialize. By 10:20 I was still sitting there not sure why I was given an appointment on a Saturday, which as per the cleaning lady was not a working day, when a gentleman in a tracksuit and running shoes walked by me, stopped, turned around and asked me if I would like a cup of coffee while I was obviously waiting for someone; I accepted his offer. He joined me with his steaming cup of American coffee, and started up a conversation that lasted almost forty minutes. We talked about art, music, literature, education in Lebanon, life in Qatar, and the booming economy of Lebanon. When I realized that I had broken my second rule with regards to appointments, I thanked him for his time and stood up to leave. He smiled and apologized for failing to introduce himself; he was the gentleman who was supposed to meet me. I was somehow confused, sat down again, turned on my business posture, and asked if he wished to start the interview. His response was: How soon can you start with Bechtel, and what is your current financial package at Orient in Qatar. Two months later I started my employment with Bechtel International in the Pipeline and Petrochemical Division, in Beirut.

Personal Interview

At that very early stage in my career I decided to adopt the same interview technique, as interviewer or interviewee: meet the other party and get to know the quality of the person before getting into work records; and treat the personal work resume' only as a document for screening purposes and selection of candidates for interview. Success and achievement are very much affected by work environment and opportunities. With the right personal attitude and positive character traits, a non-achiever could turn out to be a significant producer when properly directed and well managed.

I went back to Qatar, still hesitant whether to make the move to Beirut or not, until Salah, who by then was practically running Orient, unknowingly helped me to decide on Bechtel. Salah assumed the role of an elder brother, and accordingly believed that he had the right and responsibility to map my life and career and to mold my character to reflect his. This presumptuous attitude dramatically failed when he applied it to his own brothers and sisters, and was offensive to me even though he meant well in his own way. That Sunday in Beirut, after my social interview with Bechtel, and over a romantic dinner, Vicky and I acknowledged the inevitability of getting married. That was awareness rather than a premeditated decision, with the implementation date pending another spontaneous resolution.

The push came from Salah when he vehemently objected to my decision to get married since, in his honest yet distorted opinion, I was not professionally mature enough, and marriage would put a lid on the development of my career. On that day of the unwelcome revelation, which fell on the last day of November 1973, I submitted my resignation, an act that Salah would never forget nor forgive. *Caring for the welfare and the professional development of your employee is one thing, and assuming the right to dictate their personal life is another; the latter could never succeed as it infringes on the right of personal freedom.*

Bechtel

At eight in the morning on the fifth of January 1974 I walked into the offices of Bechtel International, located along the Airport road leading to Rafiq Al Hariri Airport, which was then known as Beirut International Airport. By the end of the day I had completed all the formalities that a new employee goes through, was assigned an office space, and was introduced to Ted Van Blomen Wanders, who was the head of the engineering department and my direct boss. The first lesson of that first day was that people at Bechtel office go by their first names. Precision is vital in the oil and petrochemical engineering fields; the Sir, Mr., Mrs. or even the family name are too general, while Walid or Ted are specific and to the point. I felt at ease with this informality and embraced it through my future career, although it was not the common practice in the business environment, and definitely not so within the Middle Eastern culture.

Design Assignment

My first work assignment was to design a centrifugal pump for an oil refinery project, for the Government of Tanzania. I was overwhelmed by such a design responsibility, dusted my engineering textbooks that I had not used for almost three years, and went about this mission in a very systematic way, starting with the design of the pump impeller. Two weeks later, Ted calls me to his office and inquires about the progress of my work. He listened patiently, seemed to be interested in the many pages that I had written and filled with all kinds of calculations, and out of nowhere he asked: "What would you have done if you required such a pump while you were working as a contractor in Qatar?" My obvious answer was that I would have picked one of the pump suppliers' catalogs and chosen a pump to suit the required service. He stood up and walked me to the technical library situated next to my office, which until that moment I had not realized was there; explained the numbering system by which catalogs, other references and manuals are filed or organized; picked one product catalog of Hayward Tyler; and in five minutes selected the pump that, after laboring for two weeks, I was still fighting with the proper alloy for the impeller shaft. He said: "Walid, we employed you for your very good hands-on experience that you acquired in the field in Qatar on the pipeline project, and that is the input we need from you. Impellers and shafts are designed by the manufacturers and you should know that since your engineering training was at Hayward Tyler Pump Division." All my worries about failing to perform, and whether I should have stayed in the construction business rather than moving to design and engineering, dissipated immediately. *I realized that in order to perform well in the engineering business, whether design, consulting, or construction, one does not have to reinvent the wheel. The secret was to know what information you need; where to look for it; how to find it; how to select the most suitable data for the required application; how to integrate it in the overall solution; and, most of all, how to do all that efficiently, on time, and at the best cost.*

From that early date my approach to my work changed completely, and mainly followed the raw instincts developed within the tough field-engineering that I acquired from pipeline welders and their foremen. At Bechtel, I was expected to approach any application from an installer's side rather than a designer's view, which I must have done well; within less than a year I was promoted to project engineer in charge of two design projects for the Government of Tanzania, in Dar El Salaam. With that promotion, in addition to financial benefits, I got what I considered the most important thing for me: my being accepted within the management circle of Bechtel Beirut. First thing I did within that circle was to convince the administration director to extend the lunch break from half an hour to one hour in order to give us enough time to finish the card game of Bridge. Since Bechtel working hours were on flexible time basis, where the employees could come and leave within 45 minutes early or late, provided they put in the required number of working hours per week, the non-Bridge-players did not complain. It was during these relaxed and informal one-hour sessions that I absorbed every technical, managerial, and business idea that happened to be the topic of the day among the high brass. *I sometimes felt that topics were raised for my benefit; later on I became aware that it was "hands-on- training" through daily interaction.*

This made me remember the discussions and arguments during late evening hours with the welders, fabricators, and foremen at the camp in Dokhan, Qatar. I had started as the assistant to the project manager on the natural gas pipeline project, and within few weeks took over the management by default after the project manager's irresolvable dispute with the Consultant. Managing the technical aspect of the work was not so difficult since I had the aptitude and the devotion for learning and had an affinity to reading; however, managing the workforce was altogether a different issue. Thanks to Tony, the General Foreman who ever since that project became a lifelong friend, I was able to maintain "Mouaawiet hair" between me and the welders. The story goes that Mouaawiet, the Omayeh Caliph, when asked how he was able to manage and control his subjects, he said: I maintain a tight hair between us, if they pull I let go, if they let go I pull. Expert pipeline welders are limited in number worldwide and move in small circles from one project to the other, always adding certificates of completion and strange stories to their personal portfolio. Their professional life is shorter than those of other professions, mainly due to the absorbed vapor from the flux of welding rods. Their salaries are much higher than other skilled technical people; their approach to discipline is temperamental; their loyalty to each other is legendry; and their envy of each other often leads to fistfights. Keeping the level of production on schedule as required by the project work program depleted my daily energy before the sun reached mid-sky. Nighttime, other than the few hours of sleep, was devoted to exchanging stories of bravado by the welders, with my frequent interjection to solve their personal problems and to explore new topics in order to improve my knowledge of the trade. The project was successfully completed; Salah had tried to impede my marriage plans; Bechtel offered me a job; and, for the following three years, I would have no direct dealings with construction manpower.

Visits to Tanzania

My first of the only two visits I made to Dar Al Salam, the Capital of Tanzania, was in February of the year 1975. Construction had started on the Tazama Tank Farm Expansion Project, and the Client had called me for a meeting to review certain design issues. The Contractor was the firm that my friend Tony from Qatar worked for, and he was instrumental in bidding for and being awarded this construction project. During my one week-stay in Tanzania I received two shocks that remain vivid in my mind to date: the car accident on my first day and the news of the upcoming war of Lebanon on my last day. At the airport in Dar El Salaam I was met by the office secretary of Bechtel and the administration manager of the Contractor, Rabah, who happened to be the brother of Salah, my friend and ex-boss in Qatar. The secretary was driving her new VW beetle car; Rabah was sitting beside her; I took the rear seat behind Rabah; and we headed towards the resort hotel where I was to stay. It was a narrow rural road; the secretary had obtained her driver's license a week earlier; the car coming from behind was approaching fast; the lorry coming towards us was barely visible with a background of the setting sun; and panic struck our lovely driver. After several flip-flops, on and off the road, the car rested upside down with the secretary and Rabah barely conscious and bleeding profusely. If it were not for the English doctor, who decided to forsake the restrictions imposed by the law on helping car accident victims, their life would have been at risk. I came away "shaken but not stirred" and a believer in the doctor's saying that it is better to take chances with the Law than with human lives.

Closure of Bechtel Offices in Lebanon

On the last day of my visit I was approached by the head of Bechtel Construction in East Africa, who was stationed in Zambia, and was asked to rejoin the construction industry as his assistant. I thanked him but refused as I had just started enjoying the lifestyle of office-engineering, and my intentions concerning Vicky were already made public and preparations for the big day where ongoing. He asked me to reconsider the offer as Bechtel was planning to close its offices in

Beirut due to the impending civil war; a statement that was later negated by the director of Bechtel-Beirut and was so ridiculous in the absence of any visible or known strife among the Lebanese people. A month later, during my second trip to Tanzania, the offer was again made and I was given one month to reconsider; again I put it aside as hallucination of an eccentric person who, being a tenior with Bechtel, took every other year off to go hunting wild animals in Africa in his custom made safari trailer. Apparently that statement was based on facts known only to those who had access to the policy makers of Bechtel International. The news of Bechtel closing down its offices in Beirut came to me on 10 August 1975, over dinner hosted by Moufeed and Meesha in honor of the newly weds, Vicky and I, the same place where two years earlier I introduced to Bechtel; that brought our month-long honeymoon to an unsettling end.

Within less than a year Lebanon was transformed from "Switzerland" of the East and the envy of the Arab World to a stateless country with a ravaging internal war that spread like a wild bushfire. Fifteen years later, the actual fighting that had involved all political parties, members of the Islam faith and the Christian faith, and the various sects within each faith, somehow stopped. There were no winners and only one looser: Lebanon. The civil war brought to light the hidden social grievances; the historical rivalry among religious factions; the frustration of the stateless Palestinians; the conflict between Western and Eastern cultures; and mostly, the social and civic immaturity among the Lebanese population. Lebanon is the site of Phoenicia, one of the oldest civilizations known to mankind, with many present Lebanese claim to be the direct descendants of the Phoenicians. It is amazing and inexplicable how the heirs of a four-thousand-year old civilization could contribute, within a relatively insignificant period of fifteen years, to destroy the basics of civilization: Coexistence, tolerance for the beliefs and opinions of others and most of all being law-abiding citizens.

Wedding Ceremony

The turmoil had started few weeks before our planned wedding day of 12 July 1975, which was a double wedding: Vicky & I and Marwan (my first cousin) & Amale. My excitement and anxiety attributed to the impending start of my second phase of my social category was overshadowed by the sudden collapse of all my past and the ambiguity of my future. Most of the wedding arrangements had to be either changed or cancelled to avoid contradiction with the chaotic rules that governed an internal war. Despite all odds, the church-wedding ceremony went well; special thanks go to our priest, whose attention to details and his insistence that the sermon be repeated twice prevented the catastrophe of marrying the four of us to each other.

Leaving Lebanon

Developments within the ten months following our marriage: Our lifelong honeymoon had started; Bechtel closed its offices in Beirut and I turned down their offer to transfer to San Francisco offices; the engineering firm that hired me to run its office in Beirut closed shop and relocated to Cairo; communication between Beirut and Monsef was severed, and between Lebanon and the outside world was often disrupted; supply of water and electricity, or the lack it, was the topic of the day; destruction was steadfastly creeping within the whole infrastructure; and, Vicky was pregnant with our first child.

Athens

It was time to leave Lebanon and find security and safety wherever we would be welcomed. The Greek embassy in Beirut was the only embassy still accessible in Ras-Beirut, and in no time granted us visit visas. With my father's blessings and my mother's encouragement, Vicky, my sister Lama, and I flew to Athens to stay with my friend Bassam and his wife Anne Marie while we looked for a more permanent country to live and work in. The three months we spent in Athens were filled with gloom, depression, worry about our families and friends back home, and concern about the future which could label us as homeless refugees. Then life shifted to auto-adjust and fate kicked-in and came up with a solution for our dilemma. My cousin Khaled, who was working in the Sultanate of Oman with Sogex, somehow found out that I was in Athens and sent me a message that Sogex had a job for me as soon as I could get there. Nelly and Nick were at that time living in Alexandria, Egypt, where Nick was running Ford Motor Company's business there, and was trying to get Ford lifted off the Arab boycott list of companies that had dealings with Israel. Nelly found out from her sister Zeineb, who had escaped with her family from Lebanon to Alexandria, that Vicky, who was then few months into her first pregnancy, and I were in Athens; we were invited to stay with them for as long as it takes to re-settle. And so, Vicky and Lama went to Egypt; I went to Oman; and Lebanon was rapidly approaching hell.

Almost four months later, in December 1976, I traveled from Oman to Beirut, via Cyprus, picked Vicky and Marwan, our 40 days old firstborn, and returned to Oman to embark on a thirteen-year employment cruise with Sogex International. We were among the first families to join Sogex and the last one to leave after the mighty Sogex ship broke down and sank in a turbulent ocean of dept and multitude of financial problems. It took us a short while to adjust to our new social and work environment; thanks to Vicky's ability to adapt without complaints and her complacent character. Instead of brooding over the hardships that we endured since the war broke in Lebanon and the ordeals of our travels till settling down in Oman, we decided to select the few funny and bright memories of that period and treat them as stepping stones along the path of life.

I will always remember the Greek taxi driver who took us back to Bassam's home from a wine festival. He had a funny looking neck that, mostly due to the excessive amount of wine we drank, was the subject of our sarcastic jokes all through the half hour drive. We felt safe and bold hiding behind the Arabic language in the Lebanese dialect. The embarrassing moment came at the end of the trip when the driver wished us a pleasant stay in Athens, in perfect Arabic, and laughingly advised us to never act foolishly on presumptions especially when it came to taxi drivers who, by trade, are famous for their intolerance to wisecracks.

Applying for extension of stay in Greece was another interesting incident that, years later, taught me a good lesson in how to behave with authorities. Our original visa to Athens, the one we had obtained in Beirut, was for one month stay only. When the deadline approached, we still had no place to go as Sogex offer and Nelly's invitation to Vicky were still in the land of the unknown. I went to the immigration office in Athens and inquired from the lady officer who was stationed behind a huge counter about the possibility of extending our stay due to dire circumstances beyond our control. She handed me a form of several pages, one copy for each of Vicky, Lama, and myself, and watched me like a hawk while I was filled them in. When I came to the space marked for stating the religion and the sect, I wrote: Christian, Greek Orthodox. The lady officer interrupted my writing; asked for the forms back; took them to her supervisor in his small enclosure; and, then asked me to go in and see him. Her unexpected action induced a sudden call of nature that I almost responded to. After a gentle welcome statement, the supervisor asked if I would like a one-year extension with the right to work as well; the call of nature receded to a manageable level. Needless to say I accepted with profound thanks and decided to attribute this generosity to my being a Greek Orthodox. Few years later, on my second trip to Greece, I made a comment to the customs official who had asked to search through my carry bag that I am Greek Orthodox, expecting a repeat of the previous preferential treatment. That was the only time in my hundreds of trips to various countries around the World that I was strip searched. Apparently not every Greek is a fanatic Greek Orthodox.

CHAPTER III: 1976 to 1978

Oman

In 1976 the Sultanate of Oman was at the very early stage of opening up to modern development. Years of traditional and conservative ruling had deprived the country of adequate and sufficient electric power, drinking water, roads, and infrastructure. Sogex International had started operation in Muscat two years earlier with the supply and construction of the first Power and Desalination Plant; the first modern hotel; the first town electrification project; and the first power transmission network. My first assignment with Sogex was the Contract Engineer on Ghubra Power & Desalination Plant; responsible for implementation of all contractual terms and conditions, monitoring subcontractors, and negotiation of contract modifications. That was the beginning of my exposure to legal and contractual matters related to the engineering and construction industry. I was captivated and there and then decided to develop my skills in contract legalities and make that field as the main stream of my future professional life; the backbone to a career in project management. Success did not come easy or cheap, and it was not all self-made. By nature, I am inquisitive, dedicated to details, meticulous, and an extremist when it comes to logical analysis. To some, those traits are found annoying; to me, they were the prerequisites to absorb and make use of all pertaining information that I could lay my hands on.

Marwan's Infancy

Going back in memory to those days in Oman, the only vivid pictures I have retained are those of the early childhood of Marwan. Like most firstborns he had to tolerate his parents' ignorance and their experimental approach to raising children. He suffered our lack of understanding of his very primitive language of ahhs and woos; we gave him the bottle when he asked for a ride in my Honda car; we put him in bed when he wanted to play; and we misinterpreted his signals for a diaper change, and ended up wasting the one we used shortly before he let it go. In hindsight I believe babies are devilish, and he did that on purpose; otherwise why is it that he smiled every time we blundered. Raising Marwan was a pleasant experience and a successful experiment. We documented our progress daily and managed to capture every aspect of his development during his first year of age. Most of the families of Sogex International were in the breeding phase of their lives, us included, providing fathers and mothers with the universal topic of: My child did that or how do you make your child do this, and always taking mental notes of every word said. The Sogex children actually contributed more than idle talk for the Sogex parents; they bonded the Sogex families through their preschool activities, birthday parties, and sports activities. This bond got stronger and stronger as the families grew in numbers and in members, and the children became the catalyst that linked the families as they dispersed or re-grouped, driven by their assignments on the various projects over four continents.

War in Lebanon

Another element that helped in bringing the Sogex families closer to each other is their shared feeling of despondence over the escalating fighting in Lebanon. Many of these families were Lebanese, a natural thing in the Arab World when the owners are of the same nationality, who could not believe how so suddenly their world turned upside down. The favored topic of after-work hours, other than their children, was the raging war in their homeland. We analyzed in depth the current events as reported over the news media or by relatives back-home when international telephone lines were not in coma. We reincarnated historic events in the region to create a linkage between past and present behavior of the West and to prove the continued interference in the affairs of the Middle East. A favorite topic was always the differences among various religions and sects which, because of their abundance in Lebanon, reflected the conflict of religions worldwide. Political parties, many of them were religious-sect oriented, had their doctrines profusely defended or criticized. Discussions got overheated and the crescendo of shouting dominated all other chit-chat, and then the call for dinner always had the effect of pouring water over fire; all that was said was forgotten by the magic taste of Kibbeh and Tabbouleh, the famous traditional meat and vegetable dishes of the Lebanese cuisine.

Lebanese Cuisine

Lebanese cuisine, food and drink, is one issue, and maybe the only one, that boasts the non-divisible and undisputable loyalty of all Lebanese people. Ethnicity, economic status, religious affiliation, educational standard, gender, and age vanish around and in the vicinity of a rich mesa. A mesa, which in Spanish means table, is in Lebanon a table covered with anywhere from 50 to 250 small boat-shaped dishes containing a large variety of hot and cold appetizers, and no two dishes are alike in content, color, taste, ingredient or name. The mesa has its own rituals and demands obedience from all participants. Relaxed posture, light talk, humming and soft singing, smoking cigarettes, cigars or shisha (Hubble-Bubble), and staying away from politics are all mandatory. And so, the wives of Sogex men took turns, individually or in groups, in preparing and setting up mesa meals to keep political discussions from reaching the boiling point. Those magnificent ladies had other reasons for indulging themselves in elaborate mesa preparation: A physical diversion from baby-care and an emotional and mental escape from their worries about their families and friends who were not so lucky to find refuge outside Lebanon.

The Family on the Move

Vicky's family and mine belonged to those Lebanese who were loyal to Lebanon and not to a Lebanese faction. They believed that the internal war was unjustified, and neither supported any specific combating group nor condoned the behavior of any of the political or religious parties. This perspective somewhat reduced the potential risks on their lives; their main worries were to secure food and medicine, and their main concerns were the availability of water and electricity in Beirut. To further escape from the reality of the ongoing destruction in Lebanon, Vicky and I diverted our minds and energy towards caring and raising Marwan. We thought at that time that the extra attention we exhibited was because we were devoted parents; however, we later realized that it was due to our inexperience in handling the total dependence on us of a helpless human being. Three years later, when we were in Jeddah, Mira was born; four years after, while residing in Yanbu, Majed was born; and with each our attention to the details of their growth faded away. Repetition is a sure way for losing interest or for taking things for granted. Being blessed with a boy then a girl and then a boy again somewhat broke the monotony of child rearing. In the same way, the diversification in my work and the several changes in our place and country of residence maintained the element of novelty in our life. We proved wrong the prevailing belief that constant move is bad for raising a stable and secure family; it bonded us, made us more tolerant towards others, polished our behavioral traits, broadened our experience, and polished our perception of life. As for career development, the exposure to various cultures and multiple professions enhanced greatly my ability to understand people and have them accept to be managed by me, even those who are more experienced or older or hold higher educational degrees.

London

September of 1977 witnessed a huge leap in the expansion of the young Sogex and lead to its reclassification from a contracting company to a large corporation. How it happened I do not know, but thanks to the excellent connections of its two owners and its chief executive officer, Sogex was awarded the largest combined power and water desalination project in the world. It was called Jeddah-IV and, as the name indicates, was located in Jeddah, Saudi Arabia, where natural oil was abundant and water and power were scarce. What must have helped in getting that project were the scary ambition of the owners and their intelligent use of the Oman project to lure the most prominent experts of their time in desalination technology. Acquisition of Envirogenics Systems Company, an old and reputable American firm specialized in water treatment and purification, was an added value.

With no time to waste, Sogex opened a huge office in London and centralized its engineering, procurement, and planning operations; all dedicated to Jeddah-IV project. I was among the first employees to join the London office, and was assigned to the *Construction Methodology Department whose function was: The evaluation of design criteria in conjunction with the latest modern construction philosophies and the most advanced material and equipment with an aim to achieve a most economical execution plan*. My role was to coordinate among the engineering, construction, and procurement departments, and was supported by professional experts in all three fields. The responsibility was overwhelming; the exposure was unparalleled; and the authority was nerve wrecking. I was still young and honestly believed that I could achieve whatever I put my mind to. I had not experienced failure, was dedicated to my work, and most importantly I had unlimited support from Vicky and her unshakable faith in me. Many years later, on our 25th wedding anniversary, I recalled an old story that I diligently tell to every newly wed couple I meet. Two close friends got married around the same time and each went his own way with his bride. They met many years later and decided to compare notes about their married lives. The first advised that he had been living a continuous honeymoon to the amazement of his friend, who had been a permanent resident in hell since his wedding day. The secret to his happiness, the first friend said, was the mutually agreed upon split in responsibilities and concerns between the husband and wife, where she takes care of the unimportant issues and he handles the important ones. To further explain to his unhappy friend, he said: My wife decides what country we live in; which house we rent; how many children we get; what car we buy; to which schools we send our children; with whom I work; what salary I ask for; where we go on vacation; whom do we visit and who is invited to our house; etc… What about you the bewildered friend asked. The reply was: I worry about the spread of HIV in the world; the increase in violence and terrorism; the shortage of wheat in Russia; and all similar major stuff. And they shall live, and so do Vicky and I, happily ever after.

It was in London where Marwan experienced most of his first independent acts, and luckily we were around to witness and record them. His first step; first time he climbed the stairs on all four; first meal without sharing the food with his clothes and the carpet; and watching his first complete football match on TV. Unfortunately, his first birthday was in Lebanon and was not a happy occasion.

My mother, who had been under treatment for cancer for several years, had passed away two days earlier. What I remember about her illness are her strength and the support of all friends and relatives in our community in Monsef. Years later, when we were living in the USA, and during a discussion about Social Security I commented that the best system in the world is the Lebanese Family System. I gave as an example the case when my mother was at the hospital and needed blood transfusion. The Blood Bank asked an employee of the hospital, who is from Monsef, for donors as was customary in such a case. Ten days later I asked for the hospital bill to settle it prior to my mother's discharge; the cashier asked me if I wanted to be paid in cash or to add the amount as credit to our account. It was then that I knew that the hospital gives monetary credit to the patient against donated blood, and in my mother's case, the credit exceeded the value of a ten-day stay, with treatment, at the American University Hospital, the most expensive one in the country.

Corporate Power

London was also the place where I experienced the power that comes from being a member of a large corporation with authority to negotiate multimillion US Dollar deals with international industrial giants of the world. The Jeddah-IV project had a value in excess of one billion USD, with many supply contracts that ran into the tens of millions. The president of a huge Japanese industrial conglomerate, who wanted to discuss his offer for one of the largest packages, invited me to go to Tokyo for negotiation. I had the audacity to tell him that I was too busy to waste my time in travel and gave him an appointment in London scheduled two days later, between the hours of 10 and 11 AM. He arrived on time to the conference room; if he were angry or upset to be in the presence of someone so much younger and infinitely less important in the business world, he was polite or restrained and did not show it. During the rapid growth and expansion of Sogex, and as I climbed the management ladder to the position of Vice-President Planning and Control, I exercised the Sogex-extended power on several occasions and enjoyed it. Until one evening in Jeddah, over a small dinner party at our house, Khaled, my good friend and ex-boss at Orient whom I had not seen for years, remarked that the rapid growth of Sogex defies logic and must be unhealthy. Sogex power, and mine by extension, are artificial and could collapse in no time; and that was something that I should be wary of. Six years later the empire of Soges was no more; *my power from then on was an extension of me only, coming from my knowledge, capabilities, and experience.*

By March 1978 Sogex was entrenched in the water desalination business through acquisition of international technologies in the field of water production. One such acquired company was Envirogenics Systems Company, a well established American research firm and a pioneer in Reverse Osmosis water purification. Envirogenics was awarded a desalination plant project for the Saudi Naval Base in Jeddah through the US Ministry of Defense. That same project was my first assignment as project manager with Sogex; my first encounter with Envirogecnics who years later sponsored my application for Green Card to the USA; and the cause for establishing my residency in Saudi Arabia that lasted almost fifteen years with intermittent stay.

CHAPTER IV: 1978 to 1984

Water Desalination

The last time I was in Jeddah was in 1993. Jeddah then was quite more advanced and developed compared to the Jeddah of 1978. I was lucky to witness this metamorphosis over a short period of fifteen years, and to participate in the booming growth of Saudi Arabia. My first assignment, in 1978, was the Jeddah Naval Base MSF Desalination Plant. MSF stands for multi-stage flash: a technique used to convert seawater to potable water based on the principle of evaporating seawater; condensing the vapor; collecting the fresh water condensate; and then treating it with mineral additives to soften its corrosiveness. This process produces large quantities of water economically, and thus saving on the energy needed to boil seawater. This is achieved by introducing hot water into huge metal boxes under vacuum, thus causing the water to boil at a lower temperature than 100 deg. Celsius (212 deg. Fahrenheit). The technique lies in balancing water and steam flows, their temperatures, and the design of the trays and tubes of these large metal boxes. The main problem was the internal corrosion of the various components which are subjected to the aggressive seawater, made more corrosive when heated. The challenges were in finding the appropriate chemical treatment systems, using acids, bases, and antifoam agents to prevent precipitation of salts and clogging of tubes, and in utilizing a combination of metals that can withstand the side-effects of these acids and chemicals. The metallurgy involving choice of metals for lining the inside of the boxes; the heat-exchanger tubes; the interfacing of various metals; and the welding material for stitching it all together was a science unto itself. In the seventies, that science was not yet fully developed and international expertise in manufacturing evaporators with complex metallurgy was scarce. I remember facing major manufacturing problems with the evaporators of the Yanbu Desalination plant, in the mid eighties, which were being fabricated in Greece. The only solution to recover the resulting delivery delays and to try to meet the project contract completion date rested in locating a second manufacturer, and splitting the work between the two. Finding a second fabricator with suitable facilities was the small part of the solution; the major issue was possession of the required skills to handle the highly specialized welding techniques. The solution was outsourcing a group of expert metallurgists and welding specialists, to join the Sogex team of specialists, and mobilizing them to Korea, to train the Korean fabricator and to supervise his work. Few years later that Korean company became an international contractor for turnkey desalination projects, competing with Sogex and the giants in the business.

Al Hamra Area

Al Hamra area, close to Al Hamra Palace, was in early 1978 at the outskirts of Jeddah city. A four-story building near Al Mukhtar supermarket in Al Hamra, with two apartments per floor, was enough to house the entire Sogex pioneering families. The offices, in an adjacent 2 story villa, where the hub from which Sogex was launched in Saudi Arabia in the construction industry, to become in a miraculously short time a leader in large housing projects, and the number one in desalination plants. The apartment building and the office block were about half an hour drive from Jeddah port and the close-by Naval Base. The roads leading to the Base were mostly narrow, crowded, and seemed to be always exploding with expatriates from all nationalities, darting in different directions with a common objective: to benefit from the disbursement of the Saudi oil wealth in return for their efforts to develop and modernize Jeddah. The work frenzy sucked us all in to such an extent that dividing the day into AM and PM became meaningless and inconsequential. It was as easy to become a workaholic as it was difficult to turn into an alcoholic; Saudi Arabia being a dry country helped in both cases.

Working Hours

While I was immersed in my work on the desalination project in the Naval Base, Vicky was forging long lasting friendships with the other ladies of the small Sogex community, and Marwan, who was not yet two years old, was sharing his toys with his peers who were mostly of the opposite sex, not that it mattered at that tender age. In no time we adjusted to a weekly time schedule of two activities: HEAVY WORK was the first activity with a duration of one hundred and forty-four hours, on a time scale from Saturday through Thursday, and LIGHT WORK as the second activity lasting 24 hours, the maximum allowed in one day by the rotation of the earth around itself. Friday, presumably the official day-off, was devoted for taking care of uncompleted office work while Marwan roamed about the project site, suspended from a hardhat with his name on it, and leaving Vicky to herself at home to catch up on important personal chores, like rest and sleep.

Safety Requirements and ILO Safety Manual

Wearing hardhats, safety boots, and safety belts was not regarded as a necessity or even essential on construction sites at those days. Safety measures were forfeited in favor of rudimentary work habits imported by a workforce of diversified cultures and hard-to-change habits. Enforcing safety practices on my project site was something that I took very seriously, especially with the added inducement of the US Army Corps of Engineers, better known there as The Corps, who were the supervising project consultant. Many years later I was invited by the International

Labor Organization to participate in the drafting of the "Safety Manual in the Construction Industry" which was later adopted and ratified by all member nations of the UN. During the discussions and debates on safety requirements and procedures and the best way to apply or impose them an interesting argument was presented by one expert who had managed many construction projects of water dams in India. His recommendation was that not all safety procedures should be applied universally, and consideration of local practices and traditional systems should be allowed. He gave a justifying example the case where an international financial institution, which had been financing a dam project, had insisted on the use of pre-designed metal scaffolding system. The local labor representative insisted that his laborers worked barefooted and were used to bamboo scaffolding. The feel of metal piping could inhibit their sense of balance especially that they climbed up and down carrying baskets filled with dirt on their head. It took two unwarranted deaths for the financer to concede and approve the use of the bamboo system. *The Safety Manual recommended that traditional local practices be reviewed and evaluated in comparison with international adopted procedures, and if found safe for use for the intended purpose they could be applied under supervision of someone well versed in these local practices.* The civil works foreman, in charge of construction of concrete foundations on the Naval Base project, complained that the delay in his progress was because of the imposed wearing of safety shoes. Many of his laborers came from the same community in a Far-Eastern country where rubber flops were the traditional footwear; safety shoes weighed them down and restricted their agile movement. Safety shoes were responsible for the daily saving of many workers from construction wood-nails; his laborers were getting preferential treatment being all brothers, as they called themselves since they came from the same twenty-thousand-square-kilometer region (twice as big as Lebanon); he was lazy and an inefficient leader; and, he got fired on the spot. Miraculously, productivity increased two folds in one week.

US Army Corps of Engineers

After thirty-five years of working for various clients, government agencies and in the private sector, and through long interaction with consultants and contractors of many nationalities, I still retain a special appreciation to my brief encounter with the US Army Corps of Engineers on the Naval Base project in Jeddah. Discipline, systems, procedures, contractual conditions, legal terms, form and formats were the rules that governed the management of any contract under the supervision of The Corps. The rigidity in their system was meant to ensure quality workmanship, cost control, and timely execution, all within a safe and hygienic environment. Most of the project managers of Sogex shied away from the Naval Base desalination project for their inability to accommodate such rigidity; maybe that was a prime reason for the CEO of Sogex to select me for the position of project manager. I prefer to attribute the appointment to my exhibited self-discipline, dedication to perfection, meticulous attention to details, and respect for rules and regulations. However, like most rigid systems, the Corps procedures could be made to work for the benefit of the Contractor, when facts indicate otherwise. The Corps contract procedure required the contracting firm to submit certain financial documents to support the percentage uplift that could be applied as cost of overhead on all additional works to a contract. Additional works are mainly works that were not within the scope of the original contract but are related and within the boundary of the project, which would be issued to the contractor as a variation order at additional price, and sometimes additional time as well. The Corps had approved a 17% company overhead uplift for Envirogenics, the percentage required by its parent company Sogex Corporation. Close to the end of the project and during the startup and testing and commissioning of the plant, the Saudi Navy asked the Corps to provide a large quantity of acids and chemicals for a two year-operation, to be delivered to the Base in one consignment. Since the cooperation agreement between the Saudi Government and the US Army Corps of Engineers did not provide for issuing Supply Contracts, which would have been the normal way to handle such a request; since only one supplier was able to meet the technical and delivery requirements of that order; and it was unhappily acknowledged that only the long reaching and significant influence of Sogex could induce it to cooperate; the Corps had no choice but to issue an instruction to Envirogenics for the supply and delivery of these chemicals as a variation order. Four hours after receiving written instructions from the Corps, I had submitted a price proposal: almost 2.5 million US Dollars for supply and delivery of material plus an additional half a million for corporate overhead and profit. The Corps supervising officer was outraged when, during negotiating the offer, I advised that the extent of involvement of Sogex Corporation was a phone call to the owner of the supply company. The US Army Corps of Engineers pre-approved 17% overhead uplift and their rigid procedures, when used to their disadvantage, allowed me to charge probably the highest cost for any recorded telephone conversation, more than four hundred thousand US Dollars.

To be a Lawyer

My thanks also go to the US Army Corps of Engineers for helping me to discover my natural affinity to the legal and contractual side of contract agreements, specifically in the fields of construction and engineering. For many years after the Jeddah Naval Base project I had contemplated going back to the university and to study law, but as an Arabic saying goes: the wind blows not as wished by the sailing ships. Work and life commitments, and maybe the lack of enough courage, kept me from following that dream. I did not become a lawyer but, by developing an in-depth knowledge of the contractual terms and conditions of FIDIC and the World Bank policies and procedures governing procurement of projects and selection of consultants, I had the opportunity to work closely with a few international law firms as an advisor on conflicts arising from disputes over contractual rights and responsibilities. FIDIC stands for Fédération Internationale Des Ingénieurs-Conseils, French for the International Federation of Consulting Engineers, and is well known in the consulting engineering industry for its work in defining codes of conduct for consulting engineers worldwide. The first such opportunity presented itself on the Naval Base desalination project, when the Corps decided to apply liquidated damages to Envirogenics for not completing the project on its contract completion date. *Liquidated damages, sometimes known as penalties, represent financial compensation to the client for the client's suffering as a consequence of not being able to use the facility, as intended in the contract, on the date specified therein. My responsibility, as the project manager, was to prove that the delay, which was a fact, was not caused by Envirogenics. There are two ways of doing that: either to prove that the delay is fully attributed to the client, in that case the Royal Navy or the Corps as its representative, or that there were concurrent delays caused by the client and Envirogenics. With concurrent delays, as long as the contract has one completion date for the full scope and not interim partial dates for parts of the project, the contractor can argue that his delay was due to his rescheduling of his work to span over the new period extended by the client's delays. Arguing is one thing and winning the argument is another. For a start, an airtight case has to be built with enough documentation to address every possible counter-argument; updated work schedules with activities logically sequenced and properly linked; supporting correspondence and minutes of meetings sorted in chronological order; and a brief of contract terms and conditions that support and justify the claim for non-responsibility for the delay and thus waiver of any penalties.*

The importance of cross referencing and tracking of documents was first instilled in me at IC, my high-school, during my tenure as chief-of-stacks of the school library. Documenting all issues related to a project, including discussions, instructions, and suggestions, by means of correspondence, minutes of meetings, and official reports, is an important management tool that was advocated by Bechtel; a technique I fully embraced and reaped its benefits on more than one occasion. One such occasion was years later when a controversy erupted between a corporate official of Sogex, responsible for overall procurement, and I, as the project manager of a major contract in Saudi Arabia, over a large order of piping that was procured and delivered to the project site with incorrect specifications. What saved my neck and resulted in the guilty party being asked to resign were notes that I had taken more than a year earlier, during a meeting presided on by that official, in which the proper technical specifications were discussed. Attendees were reminded of the details of that meeting, which were properly recorded, and were duly confirmed by them. Sogex lost a good amount of money; the official lost his job; and I gained credibility and a good bonus at the end of that year.

Penalty & Time Extension Claim

The Corps of Engineers informed me in writing, one day after the official project completion date had passed, that the contract was in default due to non-completion of the works. Work progressed; daily deductions were debited to our account; and I was fully consumed with building a claim for time extension to cover the actual delays. In a record four-week period I had put together a comprehensive report with attachments and enclosures five inches thick, and submitted it with a short and sweet covering letter, which I considered a masterpiece in ironclad claims. In twenty-four hours it was turned down by the Corps as a claim "disapproved for lacking in merit". That verdict, so sharp and short, hurt my professional image and personal ego; professionally because the claim was so tightly knit with every loophole properly covered, and personally because I had developed a flawless argument that I so strongly believed and expected others to be taken by its brilliant logic. Here the wind of the Corps decision blew in the opposite direction of my expectations, yet again proving the old Arabic saying.

First Trip to the USA

My father always said that *half the battle is won if you believe in what you are fighting for*. To win the other half I had to recruit the support of legal council to articulate the case and weigh it down with legal terms and phrases. Sogex executive management at the time believed in supporting their project managers, especially whom they had high hopes for, and gave me free hand to pursue my claim, costs not withstanding. Through Envirogenics' office in the US I hired the services of a law firm that was headed up by a retired ex-judge of the US Army Court of Appeal. In June of 1980 I made my first trip to the USA, to Washington DC, to meet with the lawyers and provide them with the documents and information they needed to build their case. After a short rest at the hotel, not that I needed much rest when physically I had enjoyed a first class seat on the long flight from Saudi Arabia, and mentally when supercharged with all the anxiety of that adventure, I decided to take a late evening stroll around the block while waiting for my hosts to pick me up for my first Main Lobster meal. The streets were so quiet and serene at 8 PM on a Saturday night, something that I was not used to for a capital city, especially that of the United States of America. Only after the way the bellboy looked at me with a mixture of horror and the are-you-stupid expression, and the later brief enlightenment I got from my hosts, did I realize the gravity of my mistake and how lucky I was to escape the inevitability of getting mugged. Apparently most of the capital cities in the world offer you one means or another to get robbed: Beirut takes your life in an indiscriminating civil war; Washington relieves you of your wallet or your life if your wallet is not thick enough; Paris dares you to cross the street and promises you a long journey between one sidewalk and the other; and some less violent cities guarantee slow death by boredom for lack of everything and anything not related to work.

Naval Base Project Claim

After six days of 12 hours of continuous meetings each day the lawyers declared that they were in possession of every detail and document they needed to complete a legal argument that cannot be refuted. A month later three copies of two volumes of legal write up were delivered to my office in Jeddah by a courier service. Even the cover letter addressed to the US Army Corps of Engineers Area Office in Jeddah and copied to the Division Office in Riyadh was included in the package. I started reading it and before reaching page 10 of 150 was so lost in its legal logic that I had to stop, consumed with multiple emotions: frustration over my inability to understand the information that I had provided them with, and the feeling of inadequacy in comprehending the structure of the logic of lawyers. All those TV shows about law firms and court cases I persistently watched had proved to be a source of entertainment only, lacking all credibility as an audio-visual brief preparatory course to become a lawyer. Upon the accept-as-is advice of the lawyers I submitted the report; this time it took the Corps one week to send it back stamped: DISAPPROVED.

In hindsight the trip to the US was not all useless and lacking excitement. A quick two-day trip to Houston to visit Munah, my Best-man whom I had not seen in more than six years, allowed me to witness firsthand a Southern hurricane and join in window taping and procurement of batteries and first-aid supplies. Another two-day trip to Vicksburg, Mississippi, where I met, for the first time, Vicky's uncles and their families, who had immigrated to the States half a century before. I was introduced to their friends who had also immigrated, or were descendants of immigrants, from 'Qornet Al Roum". My reply to their earnest questions about Tannous or his sister Mayneh or the house of Abdallah, or any of the names that I had absolutely no idea who they were, was the same: fine, in good health, and send you their sincerest regards. I hoped that none of those names belonged to people who were already dead; and if so, my interrogators were mostly too old to associate facts with time.

Sogex and Envirogenics upper management as well as the lawyers decided to call it quits; I would not throw in the towel. I requested an appointment, over the phone, with the Chief of the Corps Division Office in Riyadh to plead my case, promising that if in fifteen minutes I failed to convince him of the merits of the claim then I would stop pursuing it; he agreed. After agreeing with the Chief to withdraw the lawyers' report as it was equally confusing to me and to the Corps, I spent the remaining 12 minutes presenting a well-structured argument, organized, logically sequenced, and fully backed by the terms and conditions of the Contract between the Corps and Envirogenics. Those twelve crucial minutes condensed two weeks of an excruciating rehearsal and a summary of two years of project activities and an endless exchange of correspondence. The Chief's response raised my hopes, even though he advised against that; he called in his senior aid, a professional engineer who thought like an engineer and not like a lawyer, and asked him to sit with me for the rest of that work day, a maximum of 4 hours, and listen to my detailed explanation of the 12-minutes argument; at the end of that session, the senior had to either confirm the Corps rejection of the claim or reverse that and approve it. I am glad to say that the claim was approved and all withheld monies were subsequently released. The interesting thing is that the initial acceptance date for the project was set by the Corps as one day before the contract completion date, and a commendation letter was issued to me for completing the works ahead of time. Sogex management named me the troubleshooter of the company and rewarded me with a large financial bonus; my career as contract specialist was set then and there.

When asked about how I managed to reverse the decision of the Corps, my answer was: *the difficult part is to first convince myself and then convincing others becomes much easier. I present an argument to myself, supported with facts, and try to shoot holes in it. Weak or unsubstantiated facts are eliminated and the holes are filled with more reliable ones. Substantiating documents are collected and the cycle of argument versus counter argument is repeated. With each cycle the case gets stronger, better structured, and the facts well linked and streamlined. When my belief in the righteousness of the case becomes unshakable, I present it in a straightforward and focused manner with a sharp eye kept on the facial expressions of the members of the other party to gauge the level and the extent of converting them.* I told Mr. Taweel, the CEO of Sogex Group, that a believer can move mountains; I asked for much less from the believer in my case: don't apply penalties.

Contract Completion Dates

The Initial Acceptance Date is the first of the two completion dates that are usually set in a construction contract. It is the date on which the Client, the owner of the project, can start using the facilities of the project as intended, even though the scope of work of the contract is not completed. Uncompleted work could be work that was either not done, or was done but is not in accordance with the technical specifications of the contract, or has some deficiencies that need to be corrected. An uninstalled door to a room is work not done; a wooden doorframe instead of the required steel is wrong specifications; and a door that squeaks and has to be kicked hard to close is a deficient door. *The Client has the final say as to when he or she is willing to occupy that room in spite of uncompleted work. Of course, the Contractor is not relieved of his responsibility to complete the outstanding works; however, the Contractor would not be responsible for damages caused by the Client while occupying that room. That is called beneficial use by the owner of part of the project before the works are contractually completed. The second completion date is the Final Acceptance Date, the date on which the Contractor is considered having fulfilled his obligations under the contract, including the completion of outstanding work identified at the Initial*

Acceptance Date as well as rectifying any defects that appear between those two dates. This period between Initial and Final Acceptance dates is usually called warranty period, during which the Contractor remains legally and financially obligated towards the Client for any uncompleted or defective or deficient work. To ensure that the Contractor would honor his obligations after Initial Acceptance, the Client holds a percentage of the Contractor's earned money as Retention Money and would maintain the Performance Guarantee issued by the latter in favor of the former. In the case of the Naval Base Desalination project, Final Acceptance came two years too late due to repeated performance and operating problems with the facility, resulting from design or material deficiencies. The additional operating cost incurred by Envirogenics due to the prolonged Warranty Period was insignificant compared to the saving as a result of the non-applying of penalties.

Jubail Project

During the summer of 1980, while still looking after the Jeddah Envirogenics desalination project, I was assigned another project, as project manager, at the Naval Base in the city of Jubail, on the Eastern side of Saudi Arabia. The scope of works covered the rehabilitation of all mechanical, electrical, communication, and HVAC systems of 5000 family housing units, 13 facility buildings, and the central boilers and chillers plant supporting the base. The difficulty was not in the work itself but in carrying it out while the houses and the facilities were occupied and operational. The social habits and traditions of Saudi Arabia imposed restrictions on accessibility to the family dwellings when female members were in and the male head of the family was out at work. Planning and scheduling of the work inside the houses was a nightmare; access was restricted to 2 hours per day with activities pre-planned and detailed for 15-minutes intervals. The chief of security on the base became a daily visitor to my office to help in ironing out the flow of complaints from housewives, daughters, and mostly from the navy administrator who suffered from the absence of the navy officers and cadets, under pretext that they were safeguarding the integrity of their homes. The funny part was that most of the absentees failed to coordinate their chosen days for staying at home with the schedule of work that was pre-approved by the administrator; however, in most cases their absence coincided with some chores they needed to do around the house.

For the first few months my family remained living in Jeddah, at Sogex village, while I commuted on a weekly basis between Jeddah and Dhahran airports, taking care of my two naval bases projects in Jubail and in Jeddah. I was willing to continue my weekly travel for the sake of the stability and easy life that my family enjoyed in Jeddah; however, the airport incident in December of 1980 made me change my mind. It was Christmas Eve and I had a confirmed reservation on a flight from Dhahran for early evening, in time to reach Jeddah and be with Vicky, Marwan, and Mira just before dinner time. Marwan, my first child, was 5 years old, and Mira, my only baby girl, was 6 months old. Being with them on Christmas Eve, sharing their joy when they open their presents, and going through the hectic process of making them pause for the traditional pictures was a ritual that I would not forsake for any reason. For me, as well as for Vicky, that was a family tradition that we maintained even when we were living outside Lebanon. We all fly to Monsef for Christmas to spend the first part of the evening at my brother's house, my side of the family, and the second part of these festivities at Vicky's brother, her side of the family. With such strong family Christmas rituals there was no way I would let a trivial matter like a cancelled flight stop me from being with my family in Jeddah. Apparently, some unexpected and last minute requests that could not be refused resulted in Saudia, the national Saudi airline, having to reschedule its local flights and to cancel some. My regular flight was among those affected. Eventually, and only because of the big fuss I made at the airport, Saudi managed to fly me to Jeddah, via Bahrain, to reach my family when they were all asleep; it was past 2 in the morning. Unwrapping of presents around the beautifully decorated and brightly lit tree had happened six hours earlier.

Living in Al Khobar

The following morning, Vicky and I decided that we would never give Saudia another chance to disrupt our family traditions or to dictate our life-time schedule. Two days after the New Year celebrations were over in Jeddah, we moved to a new villa in Al-Khobar, the commercial city among the trio of the Eastern region: Al-Khobar, Dammam, and Dhahran. Dammam was the seat of all government offices and the guardian of the major Saudi seaport on the Eastern region. Dhahran was basically ARAMCO city, the Saudi oil company, and the city where the civilian cum military airport was located. At that time, ARAMCO was still partially owned by the Americans and their city, confined and contained within Dhahran, provided its residents with some extra freedom that was not tolerated outside its gates. One such luxury was the Half-Moon bay, a stretch of golden sand beach that matched the French Riviera in everything up to and down to the slimmest female bathing suits. That privilege and freedom remained extended to ARAMCO residents and their friends for years after the enforcement of segregated swimming facilities for males and females, the obligatory closure of shops, restaurants, and other commercial facilities for prayer, and the forbidding of public entertainment activities including cinemas and live musical concerts.

Basem, my friend from our Group Seven at IC who was then working in Al-Khobar, was among the close bachelor friends and relatives that we invited on Fridays, the weekly day off, for brunch and a game of cards. We noticed that for some reason his face remained serious with a non-compromising expression on most of these Friday get-togethers. It turned out that he had decided to practice non-smiling to be ready if it became suddenly forbidden to do so; that was an exaggerated act, and could have been a simple joke, but it did reflect the strict social behavior that non-Saudi residents, especially of Western or Westernized cultures, could not come to terms with. Nafhat was one of the relatives who frequented our Friday brunches. He is a cousin of Vicky and my close friend from Monsef with whom I share very many memories of those happy occasions when he drank more Arak than his brain could handle, and I sang along with him the Arabic old songs that he mastered with his rich and beautiful voice, and I followed suite with my voice that is bearable only if others are semi-conscious. Nafhat's card-playing skills were even worse than his ability to handle hard liquor, and realizing this by all those present, he was never asked to join in a game. Until one day the father of a friend, who was on a business visit, attended our gathering and, having noticed the non-participation of Nafhat, asked him to be his partner at one of those card games. Nafhat obliged out of respect for the distinguished elderly man, who had been my father's classmate at the university and a parliament ex-deputy. It took the distinguished guest less than ten minutes to lose his demure posture and, without even an apology, to demand that Nafhat be banned from even sitting close to the card-game table.

Sogex Village

Living in Al-Khobar allowed us as a family to spend more quality time together than life at Sogex village in Jeddah permitted. Sogex village was a huge compound accommodating more than 500 families and 5000 bachelors, with clubs, swimming pools, restaurants, sports facilities, and even an accredited pre-school. It had its own power generation plant, sewage treatment plant, and Reverse Osmosis water production plant. The Women Club, which Vicky served for one year as its president, handled the weekly luncheon at the families' pool, the monthly company sponsored dinner, and the galas on special occasions like Christmas party for the kids, New Year Eve, Valentine Day, and the Easter Egg Hunt. Additionally, the Club had its own ladies' activities like book reading, sewing practice, painting lessons, and few others. Sogex at that time was a prosperous corporation that could afford to pay for whatever social activities and events the ladies came up with, as long as it contributed to an air of camaraderie among the employees, and as long as these employees were willing to work at all hours, for long hours, and be on duty-call 24/7. Moving to Al-Khobar we left behind the hyper social life of Sogex village and we settled down to a calmer life in a rented villa at the CCC compound. The tennis court, which was across the street from our villa, failed to arouse in me the love for sports; however, it served as an excellent bicycle training field for Marwan. The two small training wheels were removed; I held the seat with just enough force to stabilize the bicycle; and I ran behind Marwan repeating that he should look ahead and not at the front wheel. I was the happier of the two of us for his quick mastering of a two-wheel bike as this evoked memory of my blue Hercules, the bicycle that had created a new breed of young riders in Monsef who are forever thankful for my patience and willingness to share.

Replaying in fast motion the year we lived in Al-Khobar, few events stick-out not in chronological order or because of any particular importance but because they spill over from my mind's filing compartment for reasons of their own: The two work teams I sent to help in cleaning the rubble after the riot in the holy city Mecca; the delicious one-of-a-kind palm tree date fruit that grew in a small oasis half way between Al-Khobar and Jubail and was only sold over few days along the highway; the weekly bridge game at our house with our friends the Sawayas and the Zakharays; the American-football star player, a teammate of Vicky's cousin in Vicksburg, who brought with him every time he visited us one of those delicious and mouth-watering items that were not readily available in Saudi Arabia; the Lebanese restaurant named after Maxim for allegedly serving the best steak in town, whose Chateaubriand turned out to be thinner and harder than an alligator's skin; and the time when Tina, my cousin's daughter, pointed at her diaper and proudly announced that she had her own toilet while her elder sister shared their parents restroom.

Yanbu

By July 1982 the desalination project at the Naval Base in Jeddah was still within its warranty period, the time that extends from initial acceptance to final acceptance, with Farouk, my site representative, still struggling to finalize the snag list and rectify all construction or operational deficiencies. The scope of work of the project at the Naval Base in Jubail was still not completed, with Sampath doing a great job in taking care of it. I maintained the responsibility of project manager on both projects, handling all contractual and financial matters, and relocated to Yanbu to manage the supply and construction of another desalination plant project for the Saudi Royal Commission. Yanbu, like Jubail, was being developed as an industrial city for oil based industries under the authority of the Royal Commission.

During my journey through time and two-dimensional space I have stopped at so many different places in a good number of countries. Each stop had its unique combination of inter-related characteristics that described my work and our family, with special events, some happy and some sad, some minor and some major, that together write the story of our lives. When in Muscat I was a contracts engineer on my first desalination project with Sogex, Marwan was born. In Jeddah, 3 years later, Mira was born while I was a project manager on my second desalination project. When we moved to Yanbu I was senior project manager on three desalination projects at the same time, and Majed was born. Other than the fact that the projects I was involved with at these three stops had to do with water desalination, for the same Sogex Group, each stop had its own small story to tell. A philosophical question: Is a person's life one story of many chapters, or many small consecutive stories linked by time, or no story at all but a series of events that might or might not be related? I have no answer and I believe that my memories of my journey so far are surfacing in no particular order and follow a pattern of their own. What I am doing is putting them in words and sorting them in a flexible chronological order to avoid a traffic jam of chaotic sentences.

Connections in High Places

In business it is a must to have good connections at high places, especially among the decision makers who bear an influence on your work. A good connection is a good relationship that could be based on one or more elements, where each party to this relationship derives a benefit from it. A common friend, family ties, previous employment, membership in professional or social societies, and just mutual respect, are among these elements. The benefits could be materialistic, monetary, exchange of services, improving on social placement, enjoyment of common interests in sports or art, intellectual excitement, and even provision of relaxing divergence from daily headaches. At the Royal Commission in Yanbu I had started with two such connections, and later cultivated a third one. David, the Director of the Prime Contractor who managed the whole development of the Yanbu Industrial City on behalf of the Royal Commission, was my first connection. David's prior employer for a long time was Envirogenics. The second connection was Abdel Wahab, the director of facilities, a Saudi national who at one time had represented his Government at the International Labor Organization (ILO), where my father, who was on the board of directors of the ILO, had acted as his mentor. Abdel Wahab was a member of an old and prominent family of Jeddah who boasted that his father had handed over the key for the city of Jeddah to King Abdul Aziz, the founder of the Kingdom of Saudi Arabia. Our friendship started soon after we moved to Jeddah in early 1978, where he was an executive of the Chamber of Commerce, and within a short time it was extended to our families. He had a progressive mind with a sharp tongue and a distinctive intelligent sense of humor. Our frequent encounters, other than when the families were exchanging gifts over Christmas or Ramadan, were duels of wits with indifference as to who comes out the winner.

With Abdel Wahab as a common friend, the third connection was the Director of Finance with whom a friendship gradually developed but was restricted to a business relationship. He helped in speeding up the processing of my project's monthly payment applications, and in return I gave him technical advice on engineering and contractual issues, whenever he required an impartial opinion, to resolve a conflict with other contractors. It was he who told me this little meaningful story when I asked him to help in easing up the tight cash flow of the project:

A Director of a company called in his Engineering Manager and asked him how much is 2 plus 2. Suspecting some deep connotation for this simple question he spent two hours researching through his technical library, but the only answer he could find was 4. When the Administration Manager was asked the same question, and after hours of reviewing all the company manuals and procedures, he gave the same answer of 4. The Financial Manager took a full week to come up with the same answer. However, when the same question was put to the Legal Manager his response was instantaneous: how much would you like the sum to be?

Though he was the Director of Finance, he decided to give me the Legal Manager's support when I left Yanbu in June 1984 and moved to Paris as the Sogex Group Vice-President for Corporate Planning and Controls. He approved the project's payment certificate reflecting 93% completion on billing basis, and payment was received accordingly, when, as the Consultant argued, the project was only 81% complete if assessed on actual physical basis. Sogex owners were happy as they had collected the project money upfront; I was happy to relieve the project's tight situation and meet its financial commitment; only my construction manager was unhappy as he had to finish 19% of the contract works with very little and non-proportionate revenue from the Royal Commission, and minimal financial support from the Group Corporate coffers.

Few years later, when Sogex was crippled with its financial burdens, I realized that *upfront payment in the contracting business is a temporary relief of a company's financial difficulties and not a cure. It is a sword with two edges.*

Unless the root issues are properly managed and resolved, the company would eventually succumb to these financial burdens; the answer of the Financial Manager would be more befitting than that of the Legal Manager.

Scheduling and Network Programs

The variance between progress based on billing and progress based on cost was well understood by Sogex management. It starts with a scheduling and work planning exercise which, when approved by the parties to any project, produces the work program that monitors the execution of works, the ordering of material, the manpower allocation, and the payment plan of that project. First, a list is made of all the activities that need to be accomplished to execute the project, from start to completion. These activities would include engineering and documentation; material procurement and delivery; mobilization to site and demobilization; establishing of offices, workshops, and other on-site facilities; execution of work; and testing and commissioning of the facility. Then these activities are linked in a proper sequence within a logical network. You cannot paint a room before you buy the paint, and you need to excavate a trench before you can backfill it. *A time estimate, or duration, is assigned to each activity by experienced people in the related field of work, who usually have a databank of related information. A lot of this information is published by authorities in the construction industry; however, each company uses data based on its past performance. The activities are then resourced; manpower is assigned by profession and number of working days, material for each activity is identified, and the necessary equipment is reflected. Based on established costs of these resources, each activity would develop a price. Using any of the commercially available computer software programs that are designed for project planning, the durations and resources of the activities are added to arrive at a final project cost and a date for completing its scope of work. Since a project usually has a restricted time period for its completion; the contractor has a pre-estimated budgeted cost for the works; and manpower resources are limited, the network logic, the durations of the activities and the resources are modified several times before arriving at a satisfactory program with optimum resources utilization and cost estimates. Logic, experience, construction management skills, and cost awareness are essential characteristics of a planning and scheduling expert.* It takes one painter one 8-hour day to paint one wall of a room; four days to finish the room; two days if you put 2 painters to work in parallel; how many painters to finish the room in half an hour? The answer is 64 painters, provided they fit into the room, can work every 16 of them at each wall, and a special paint is invented that can be applied permanently and in sufficient coats in half an hour.

A well-resourced work schedule with proper costing of its activities would reflect overall percentage completion relative to the total incurred cost against these activities. However, since contractors tend to be secretive about their actual costs, mainly to keep their profits unknown to other parties; since it is much easier for clients and contractors to measure progress by directly estimating the physical completion status of an activity; activities are given monetary values, mostly derived from the contract priced bill of quantities, with the sum of these values adding up to the full contract value. At the end of every interim reporting period the percentage completion of every activity is assessed; these progress percentages are converted to money; and the total sum would be the billing for that period. Overall project progress is the ratio of total billing to the total project cost.

By the time I left Yanbu and moved to Paris, most of the activities on the desalination project were nearing completion, such that with a little stretch of the imagination and a client who was not too strict, the percentage completion of each activity was only slightly overestimated.

Life in Yanbu

The Industrial City of Yanbu, which was under the control of the Royal Commission, was on the outskirts of the old city of Yanbu. It was almost self-sufficient catering for all the needs of its residents. If it were not for the long sandy beaches of Yanbu city, where we spent many of our Fridays with friends and their families, or for the local airport from which I took regular weekly flights to follow up on my project in Jubail, we would not have visited Yanbu city in the two years we were there.

The Industrial City had several housing communities, called camps, which were restricted to individuals and families working on various projects of the Royal Commission or at the service facilities related to the Industrial City. There was a light industry park; camps for families of senior staff with individual villas and common recreation areas, including swimming pools, restaurants, and shopping malls; camps for junior staff with multi-story buildings and their own recreation and service facilities; and there were the labor camps. Since socializing in Saudi Arabia was almost restricted to within the boundaries of one's house, except for occasional meals at the few and mediocre restaurants, additional effort was made to make the in-house environment as pleasant as possible. The Royal Commission took that into consideration when they furnished the junior staff apartments, paying special attention to the kitchen and its equipment to the dismay of my friend Pierre. Pierre's usual Friday morning started with a tennis game and was then followed by a breakfast of eggs and beef bacon that he prepared in his modern kitchen. What he did not bargain for was to doze off while the eggs, swimming in a pan full of olive oil, started burning; the exhaust hood caught fire; the fire alarm shrieked as it was supposed to; and Pierre came back from dreamland not yet aware that his uneaten breakfast was, at the end of that month, going to cost him almost $2000 to pay for damages.

The closest family within our small social circle in Yanbu was that of John and Aida. Their children were of the same ages as Marwan and Mira and were classmates at the International School. Majed and their youngest daughter were both born during our two-year stay in Yanbu. While Vicky and Aida sat around the family pool trying to keep the two toddlers cool and calm, not so an easy task in a place which is hot and humid most of the year, their elder siblings jumped in and out the water enjoying every wet moment of their time. John and I joined these family outings in the late afternoons and on those Fridays that we had no energy to drive to the beach, even though a common friend, who was a manager of a large catering company, took care of food, soft drink, and setting up a large tent at the beach with staff to provide full service. Providing such services was not uncommon for managers of companies in Saudi Arabia at the time. The five-year development plans were in full force and supply of managers, Arabic speaking and Western trained, was less than the demand. Qader, my company provided driver, drove the children to and from school, to and from their various activities; drove Vicky and her friends to and from the shopping mall; and baby-sat for Majed whenever Mathieu was busy. I did see Qader occasionally when he drove me to and from Yanbu airport if it so happens that he was in Yanbu and not with Vicky and the children in Jeddah, a three-hour drive that they frequently endured to visit with their many friends at Sogex Village. Mathieu was the family morning houseboy, taking care of the daily house chores, and the afternoon office-boy handling mail and telexes. In 1982 telexes were more common than faxes, and mail was a paper medium that popped out of clicking typewriters. Desktop computers were scarce; mainframes were still in fashion; and Kaypro, maybe the first portable computer, was still unheard of. A year later, Avie bought me from California a $5500 Kaypro-10 that looked like a sewing machine, with a cover that flipped open to become a keyboard and to reveal a tiny green screen, which served more as a status symbol than a useful tool. Still, I carried it around in its special carry case when we relocated to Paris in 1984; to Corona, USA, in 1989 when we went there to reside as immigrants; and to the ex-Kaypro technician in Los Angeles in 1990, who ended up buying it for $100 to be used for spare parts rather than repairing it at a cost of almost as much as buying a 100 times more powerful cloned desktop.

Employees

Mathieu was an example of the Indian university graduates who flocked to Saudi Arabia and the rest of the Gulf region seeking far higher income than was available in their own home country. They endured jobs that required less technical skills or experience or education than they had in return for saving enough money to build a home, get married, buy a farm, or any combination of these. He was ambitious and wanted more than money, which between his salary, overtime, and the bonus he got from Vicky for his dedication to the children, was more than he had bargained for. Mathieu had known Joseph, our houseboy in Jeddah who was miserable with the boy-suffix and wanted so badly to change his occupation to something that better fitted or better reflected his extra-large size. Joseph chose to become a pipeline welder; I authorized two months of daily afternoon training with the best welder I had on the desalination project at the Naval Base; by the end of 1979 he was the only welder to qualify for stainless steel welding and his company job-card had months earlier been changed to read welder rather than houseboy. Mathieu wanted to be an office clerk and got his chance when Vicky and the children went to Lebanon for the summer vacation in 1983. His daily work during that summer started with a two-hour house cleaning and then followed by eight hours of assisting Raju, my secretary who, like Qader and Mathieu, gave me and my family indispensable support all through my years with Sogex. Long before the Saudi Government took over Sogex operations and projects in December of 1988, Mathieu had enjoyed the position and the financial benefits of an assistant administration manager through successive promotions, which he well deserved and worked hard for.

The extent of services that Sogex and other large companies provided their employees, especially the senior ones, is what made life bearable within a restricted society and conservative social codes of conduct. However, the motive was always financial, in absolute value as well as in comparison to markets outside the oil-rich Arab countries. While living in Yanbu, my project staff and technicians took care of every small chore, which we took for granted, and did not realize how pampered we were until we moved to Paris. The electrician changed a burnt bulb; the carpenter oiled the squeaking hinges of the door; the mechanic filled my car with gas; the driver took the children to school; the laborers carried and moved furniture around the house even the light pieces that a ten-year old could carry; and the technician installed my music set and the TV, not because it was difficult for me to do so, but because it was expected from the support system of the company. The employee who installed my music set when we first relocated to Yanbu came to my office one late afternoon asking for advice: "Sir, you have a professional music set so you must know about what components to buy. Can you advise me as I want to buy a good one even if it is expensive?" I asked him what kind of music he listened to and enjoyed. He said the 777 from the Oriental music shop; as I found out, these were the cassettes which were recorded locally from a copy of a copy of a copy of pirated Arabic songs by third rate singers. I advised him to buy a small Chinese-made radio-cassette which went for 5 to 10 US Dollars. The most difficult part for me was to dress my advice in a sincere tone, with convincing justifications, without bursting the bubble of his pride. *The basic idea is to always choose the tool that best fits the intended or required output; over-specifications are a waste of money and under-specifications result in a product of poor quality.*

CHAPTER V: 1984 to 1989

Moving to Paris

Moving to Paris was a turning point in our life; family, work, and social. So many things happened during those two years we spent there, with all these events fusing into an arrow pointing us towards immigration to the USA. As years passed, the only pre-Paris thing we missed was the organized life we enjoyed and the full service we came to expect from our employees. In those days very few countries and not many companies quickly responded and fully catered to all your needs, including changing the electric bulb in your home. The Paris and post-Parris eras started with a phone call I got from the Chairman of the Board and part owner of Sogex advising me, in a matter-of-fact tone, that effective that day I was the Vice President Planning & Controls for the Sogex International Corporation. I was to relocate to Paris effective immediately and had one week to settle down before reporting to his office to discuss my mandate. That was neither a surprise nor a difficult timetable to comply with. Few phone calls to the administration departments of the company in Jeddah and Paris, some by my secretary and few by me, got the ball rolling. Four days later all arrangements for our travel, shipping of my car and our personal belongings, locating an apartment for us in Paris, and readying my office with the required support staff were completed. During those same four days I met with the Clients' representatives on the Jeddah, Jubail, and Yanbu projects and coordinated with them the transfer of responsibilities to my acting project managers on all three projects; gave last minute instructions and directives to my staff; ensured that the dedicated trio: Qader, Mathieu, and Raju are well taken care of; attended farewell parties held in our honor; and arranged schooling for Marwan and Mira at the International School of Paris.

Paris Apartment

We were not strangers to Paris, neither culturally nor to its geography, and we spoke the language. After basking in the comfort and luxury of a suite at the Meridian hotel for two weeks, we moved to a furnished apartment at Rue Franqueville, owned by an aristocratic French duke and registered in the name of his daughter. Calling it a furnished apartment is significantly belittling it because of its huge size, expensive antique furniture, magnificent paintings, and stature of its owners. The building was square with a courtyard in the middle; four floors each consisted of two L-shape apartments linked at their extremities; a basement with wine sellers and storage facilities; and the top floor dedicated to servants' quarters. The huge ornamental Iron Gate that transferred you from the sidewalk to the inner courtyard had a series of brass plaques with the names of the occupants richly engraved: Baron so and so, Duke and Duchess of somewhere, and one English Lord. The plaque of apartment 4-south stayed without a name for the two years we occupied it since we were the only tenant; the lawyer in charge of the building could not assign an appropriate title for me to match the prominence of the other owners; and the concierge was not convinced that I was the Duke of Monsef, either because he did not know where Monsef was, or my cousin Marwan was not convincing enough when he asked him for the apartment number of his Excellency, Mr. Walid Nasr, the Duke of Monsef.

In the Western world people were so used to plastic cards that a sudden appearance of large sums of cash caused a state of confusion and bewilderment.

That was how the real-estate agent and the Duke's daughter reacted when I opened my briefcase to look for the documents that they absolutely needed to action the lease agreement, which I knew of and did not have, and saw stacks of French Francs in rubber-bound bundles. In less than half an hour on that Saturday afternoon I was out of the agent's office with the lease contract duly signed and witnessed and the keys of the apartment in my pocket, leaving behind two overwhelmed French citizens with almost one hundred and ten thousand Francs, the equivalence of 4-months rent, wondering what to do with the money until Monday morning when the banks reopen for business. Displaying cash in that manner was not initially intended to achieve the unachievable but was mainly a habit acquired in Saudi Arabia. There, it was not unusual to walk into a car dealership carrying a bag full of cash to buy one or more cars. It was normal to walk into a bank and cash a check for fifty thousand Saudi Rials, equivalent to almost fifteen thousand Dollars, and walk out without counting the money; if you hold the progress of the queue to count the money you would be sneered at by other customers and asked, not so patiently, to move on by the cashier. If when you get home you count the money, as any normal human being is expected to do, and find a missing hundred Rial bill, you attribute that to God's will and write it off as a non-claimable loss.

The required documents that I was supposed to have for renting the apartment were an employment contract with a French company and a residence permit. I had none of these at the time, and unfortunately I never obtained them to my great loss as it turned out few years later. The French management company within Sogex International Group, which was supposed to be my employer in France for legal purposes, had some administrative conflicts with the Labor Union when I first moved to Paris. A freeze on hiring or firing of staff was imposed by the Labor Court. Since my family and I had a residence permit in Saudi Arabia, it was easy and quick to get them a non-working residence permit in France and a multiple entry visa for me, renewable yearly. Government formalities could then be taken care of by Vicky; the children could go to school in Paris; and I was anyways traveling in and out of France, on a monthly basis, and could obtain a visit visa to any country from Paris as well as from Saudi Arabia.

Car Registration

The only complication I had in Paris which is worth remembering, due to lack of a residence status, was with my personal car that I had shipped from Saudi Arabia. The car was shipped, along with our personal effects in the same 40-ft container, a month after we arrived in Paris. The clearing agent at the free zone area in Paris asked to see our passports to verify our legal status in country, and requested a letter from the company stating that the Persian carpets and other valuable items are for our personal use and will be re-shipped out of the country; otherwise, we would have to pay import taxes for them. We complied; no tariffs were levied; the personal items were transported to our apartment; and my car was cleared and parked in the free zone warehouse pending registration with the police and obtaining license plates. The complication started at the police station where I was informed that the only way to register a personal imported car is if I had owned it for more than a year in the country from where it is imported, and secondly I must have a residence permit in France; otherwise, I would have to pay import taxes. I had owned it for more than a year in Saudi Arabia but had no residency in France. The alternative was to register in the name of Vicky who had a residence permit, but could not get her a proof of ownership from Saudi Arabia since women there were not allowed by law to drive and so car dealers would not sell cars to non-Saudi ladies. My proposed solution to go back to the customs authorities and pay import taxes was refused since the officer in charge, after checking Vicky's passport, had assumed that I had a residence permit as well and failed to check my passport before processing the import papers with tax exemption. So, my car was in the warehouse for three months; I could not drive it out because it had no license plates; and, I could not re-export it without exposing the customs officer and his mistake. The solution came from a Lebanese acquaintance who suggested that I just drive the car out and away, since its papers showed that it cleared customs in a legal manner, and go to the nearest gas station and have license plates made with the Saudi license plate number. I did that, and the gas station attendant was satisfied with copy of my passport and the Saudi export documents that proved my ownership, legal import, and the Saudi license number. During the first year we lived in Paris I drove the car all over France and was never stopped and asked for car registration papers, except on two occasions, near our house, by the security of the OECD. In both cases they saw the Saudi car papers which were authenticated by the French embassy in Jeddah and my passport with the stamped Saudi residence permit, and assumed that I was a member of the Saudi diplomatic mission, since the Saudi Ambassador's residence was across the street from mine. I had no incentive or inclination or the stupidity to correct their assumption. At the end of the first year the law permitted that I sell the car to Vicky, who in turn got it registered in her name, with new proper license plates, with no questions asked, and no objection on my part. A year later I resold it prior to leaving France at a price much higher that what I bought it for.

Duke and Duchess

We did not see much of the Duchess who stayed mostly in their chateau in Normandy. However, the Duke, a distinguished elderly gentleman, used to pass by whenever he was visiting in Paris to share with us a cup of Turkish coffee and a bite of the special homemade cakes that Vicky liked to make and I loved to eat. On few of these occasions the Duke brought with him antique objects from his chateau until gradually our apartment looked like a small museum. Two such pieces are well documented in our photo collection: the bed cover which was a wedding gift from President Charles De Gaulle's mother, the Duke's Godmother, and the Aubusson tapestry which Vicky's father, Uncle Roosevelt, loved to sit in front of it during his visit to Paris, and patiently tolerated Majed's jumping on and off his lap while he told him made-up stories about all the pictures woven into it. Friends and relatives who visited us for the first time treated this magnificent tapestry the way tourists treated the Mona Lisa at the Louvre Museum: they made sure to have their photos taken with it.

Touring France

The tourist's fever stayed with Vicky and me during our whole stay in Paris. The few days every month that I was not on business travel around the world were mostly spent visiting every attraction, monument, chateau, museum, and restaurant, time not withstanding. When our children had school holidays, and these are frequent in France, we took long trips lasting few days each to Normandy, the French Riviera, the Loire Valley, and the various ski resorts in the Swiss Alps. It was on these special trips that we spent quality time as a family, and on the long drives to and from them I invented verbal games to keep the children's impatience within tolerance. "Pants and Shirts" is the name of the game where I gave them a word and they had to give me a matching one. We alternated among English, French, and Arabic languages to make it a game cum educational exercise. "My Boat is Transporting" so and so is a game where goods that could be transported by a ship were named, all starting with the same preselected letter of the alphabet, and he or she who repeats a word or gives a non-shippable item loses. The rules were occasionally bent to allow Mira and Marwan to be alternate winners. Majed was too young to participate; Vicky lost quickly and intentionally to maximize the bonding between my kids and myself; and I would not admit defeat and so created words that were not yet invented, at least in the three languages that we knew.

Each trip has added an array of memories that are mostly documented by photos, videos, or filed in our mental data banks to be retrieved when nostalgia hits us hard. It was in Megeve that our car failed to make it through a snow blizzard to the top-of-the mountain restaurant, the Sauvignon, and had to resort to a horse driven sleigh, with the five of us huddled together and wrapped in sheepskin fur; the Couscous was worth every second of it. Visits to Megeve or to Chamonix or to other winter resorts were countered by summer trips to Cannes, Nice, Juan-les-Pins, and Golf Juan, with occasional hop-over to Monte Carlo.

The Blue Beach at Golf Juan was where Marwan and Mira and their cousins Rami and Carine had their first exposure to topless female sunbathers. The children were still of tender age and, despite the fact that they had had constant exposure to Western cultures, they were apparently under the influence of Middle Eastern cultural behavior; public exposure of certain parts of the body was tantamount to funny acts conducive of hysterical laughter. My cousin Marwan and I tried to act chivalrous in the face of these modern aspects of civilization, but I don't believe we were very convincing, neither to Vicky and Amale nor to our children.

My Father's Second Marriage

Visiting Nice, although equally enjoyable, had a streak of strain routed in our conservative Lebanese culture. My father, after almost ten years of being a widower decided to marry Catherine whom he knew from the ILO conferences. She had been a good companion for him, something that he missed especially after each of us, his children, were living away from Lebanon and in different parts of the World. Amin and his family were in Saudi Arabia, my family in Paris, and Lama was residing then in Geneva. The combination of being alone in Lebanon and his severe disappointment at the systematic destruction of the Lebanon that he was unconditionally loyal to and had helped in building its institutions, were quickly driving him to an overwhelming despair. The straw that broke the camel's back was the breaking in and the occupation of his home in Beirut in June of 1985. The orders to the squatters came from a political entity that had failed to replace my father with its own appointee to the executive board of the International Labor Organization. That position, reserved to the Employers Group of Asia, was held by my father for many years and was internationally prestigious enough to warrant fighting for. Adding insult to injury, the apartment was taken over when my father was in Geneva representing the government of Lebanon, as a diplomat, during the ILO General Assembly conference held each June. Forty days later, through interventions at the highest political levels in Lebanon, the squatters evacuated the apartment and cleaned it out almost completely. The only things left behind were large wooden pieces of antique furniture that were too heavy to carry, or seemed too old and out of taste for the occupiers. Another thing was left behind: the large collection of books, some very rare and expensive, which we attributed to the squatters avoiding contamination by words of literature and poetry. As soon as Vicky returned to Paris, after taking charge of moving the remains of the apartment to Monsef, my father turned in the key to the landlord, sold his Buick car with the famous and well recognized plate number 646 at a ridiculous low price, and left to Nice. He married Catherine and did not return to Lebanon until ten years later, when his age required that someone take care of him, and his wife's health did not permit her to do so.

My father asked Lama and Amin to meet him in Paris, when he was visiting us from Nice, to convene a family reunion and share our thoughts and comments on what appeared an already made decision to remarry. We debated the pros & cons, and while emotionally we found it difficult to accept that another woman had taken the place of our mother in his heart, we were sensible enough to realize that remarrying was good for him. He settled in Nice and we took care of his living expenses, since he had been on retirement for a year or so, and had already transferred the ownership of the family house in Monsef to Amin, and the few lots of land to Lama and me. Marwan, who was ten years old, did not appreciate the intrusion of a new family member in his grandfather's life, but his quiet and reflective nature prohibited him from expressing this openly. My father, Marwan Senior, could sense that subtle non-acceptance, especially that Marwan the Third, our son, as well as Marwan Junior, my first cousin, had a very similar character. After his divorce and moving back to Lebanon, my father tried to make it up to Marwan the Third, but unfortunately neither of them lived long enough to catch up on lost time.

Raymond Edde

Among the things that we enjoyed immensely when we visited Nice were the political discussions we had, over dinner at my father's, with Mr. Raymond Edde. Al Ameed, as he was called, was a pillar among the Lebanese veteran politicians and an expert with international law. During the civil war in Lebanon he was targeted for his political views, openly and boldly expressed and carefully listened to by a large sector of the Lebanese people, and his life was in jeopardy. He was evacuated to France, after several attempts on his life, where he used the media and his circles of discussions as a forum for political analysis and predictions that mostly came true. My father did not always agree with Al Ameed on his political affiliations however, they shared a deep knowledge of social and political history and sound analytical minds to deduce future acts from past behavioral trends. They were not taken by surprise by the eruption of war in Lebanon and had both anticipated it. During the economic boom in Lebanon in the fifties and sixties I remember my father repeatedly saying: How can a society maintain sustainable growth and development when the citizens refused to stop at a red traffic light, and the streets were the wastebaskets for the motorists as well as the apartment dwellers. Till this date, after a civil war that lasted more than 15 years, the few traffic lights that are still standing are mocked, and the streets retain their secondary function as wastebaskets.

Paris Sites

Touring the provinces of France complemented our almost daily excursions in Paris, with an overpowering drive to discover its beauty and experience the Parisian Life. At the famous seafood restaurant Chez Francice, Majed, not yet 2 years old, bit a piece off the ultra-thin glass cup when he insisted on drinking water by himself. Rue de Passy where Vicky lost a ring and 2 days later, on our way to the food market, we found the ring wedged between the street and the sidewalk. Jardin de Ranelagh, the park next to our building which Marwan and Mira used to cross daily to go to school, is where our children spent most afternoons chasing a football or running around and enjoying it. Avenue des Champs Elysees boasted the first American fast-food restaurant which we visited every Saturday for its famous hamburgers, despite the protests of our French friends who considered fast-food an insult to French traditions and culture. Bois de Boulogne, also within walking distance from our place, is where ducks and birds are fed during the day and sexual fantasies come true at night. We took Vicky's parents for a night tour of the Bois to feed their curiosity; Aunt Georgette was shocked and wanted to cut the tour short; Uncle Roosevelt was disturbed but he suggested we repeat the tour a second and a third time to make sure that what he was seeing was for real and not figments of night shadows. His quest for knowledge and truth prevailed.

French Cuisine

French cuisine and French wine are food for the soul more than a need for the body. What to eat, when, and where, and selecting the partnering proper wine and proper vintage is an art that is acquired by experiment and systematic coverage of the restaurants listed in various gourmet guides. When in Paris you live this experience; when not in Paris you read about it, memorize names and ingredients, and try to impress others with a knowledge that you most probably do not possess. Since we believed that eating is a pleasure and not a necessity, we elected to expand this pleasure by covering as many restaurants as possible. Three years after leaving Paris we were still struggling with the extra weight that reflected our pursuing of knowledge, despite our diligent conformance to most of the diet regimes prescribed by friends and foes alike.

Sogex Paris Head Office

At the time that I relocated to the corporate office of Sogex in Paris, the Group was facing financial problems. Many projects, especially the huge ones in Saudi Arabia, were nearing completion, and were at the stage where incoming cash from billings was much less than the required expenditure to complete the works. It is well known that *at the early stages of a construction project, assessing and quantifying the monthly work progress is subject to approximation. The extent of this approximation and whether it is made on the plus side or the negative side depends largely on the relationship between the Client and the Contractor. After all, contract procedures allow interim overpayment or underpayment as long as the measured work is reasonably quantified and the estimated percentage completion is close to reality. As the project nears completion, quantities of work are more accurately measured; a closer look is taken at the quality of the presumably finished work to identify corrective actions or required replacement; and the Client tries to insure that enough money is withheld from the Contractor's payments to cover for all the outstanding work. It is a common knowledge in the construction industry that the cost incurred to move from 90% completion to 100% completion usually exceeds the remaining money to be billed and received. The definite outstanding receivable sum is fixed and is the difference between the contract value and the actual amounts paid. Potential receivables are claims submitted by the Contractor for what he believes are extra works, for which the Client is still to accept and approve. The cost to complete is elastic; the longer it takes to complete this last ten percent the higher the cost would be, and more so if new defects arise at this last stage. Ideally, the cash plan of the project should accommodate this negative flow at the end of the project, and the Contractor should restrain himself from over interim billing despite the temptation of upfront payment.*

Performance Auditing

As Vice-President Planning and Controls, my first mission was fact-finding to determine the actual status of all Sogex Group projects and operations. That was a pre-requisite to establishing a long term development plan and a short term action plan, then defining the monitoring and control mechanisms to achieve such plans. My first decision was to disregard all routinely generated documents, in-flowing from Sogex offices worldwide, such as progress reports, financial statements, and human resources data, and to carry out a complete performance auditing using my own staff. The Chairman of the Group issued the open-drawer-policy directive, circulated to all directors, general managers, and project managers, giving me the authority to request, and if necessary demand, any and all information that was pertaining to each Sogex entity. It took a year to accumulate and assimilate the huge amount of information that was either retrieved from archives or from current records. Half way through that period and for a few months beyond it, the filtering operation went on to separate facts from fiction; over or under estimates from accurate numbers and figures; over ambitious projections from more realistic expectations; true from false data; and most important the honest and reliable sources of information from the deceitful or ignorant ones.

Almost two years after my move to Paris, I had completed the first draft of a performance-audit report that better reflected the financial status of Sogex Group, the path along which its income and non-income generating operations were moving, and the future to which it was heading. On all accounts the findings were more than discouraging. Financially, the Group was in the negative especially when the financial status is measured at the completion date of its ongoing projects and commitments. The main reason for the financial problem was that potential receivables from claims for extra work on some of the mega-projects, which were taken into consideration in the Group's accounting system, did not materialize. The resulting deficit alarmed the many banks that had extended facilities to the Group causing them to apply the brake on further support. The road along which the projects were moving was bumpy, full of obstacles, with blind spots and unclear bends and curves. Clients were pushing for completion of the last 10% of the projects, when defects and snag items were coming into light and the resources were dwindling fast. The management and senior staff of these projects were like runners in a long-distance race who reach the end of the track, at their last breath and drop of energy, to find a Keep-Going sign instead of the End sign. Many of them just gave up and quit. The path was fast leading to a bleak future and quick action was needed to try to resuscitate the Group.

Liquidating Sogex

A recovery plan was quickly designed with a three-tier support system: cash injection; sale of major assets; and extended facility by the banks collateralized by commitments from the Governments of the countries where the projects were located. The recovery plan, which took a year to put in place and activate, saved the projects, the creditors and suppliers, and more than ten thousand employees. The winners were: Clients of the various projects who saw their projects to completion; the Governments who did not have to go through a long process of reengaging contractors to complete the work; the banks which avoided the calling-in of the bonds, guarantees, and securities issued by them that totaled hundreds of millions of dollars; and the employees who received their outstanding salaries and end-of-service compensations. The loser was Sogex Group, which for few short years held the number one position among Arab-owned international construction companies. It came to a face-saving end: dissolving the Group without declaring bankruptcy.

In September of 1986 the owners of Sogex Group decided to relocate to Riyadh, Saudi Arabia, from their ultra-luxurious life style in Paris, to be on location of their largest financial deficit, in a last effort to salvage the Group and retain their ownership. I, the unfortunate fact-finder and the only senior employee who was still a resident of Saudi Arabia, moved back with them, my family in accompaniment. By the summer of the following year the owners gave up on retaining any part of Sogex; my family left Saudi Arabia for good and moved to Lebanon; and I stayed behind to help in finalizing the recovery plan. I had another strong personal motive for staying in Riyadh which had to do with settling my account with Sogex.

To safeguard the interests of Sogex employees in Saudi Arabia, the recovery plan assigned them to the various projects that were taken over by the beneficiaries under the Government's direct control. As a vice-president of Sogex, despite being a resident of Saudi Arabia, I was considered a member of the corporate office in Paris and accordingly was excluded from the list of employees benefiting under the recovery plan. As luck, or the lack of it, would have it, I was the only corporate officer who was not a registered employee in France and thus my interests were not protected by the French law. By May 1989 the recovery plan was in full force; my chances of collecting my entitlements were less than nil; I had written off the money owed to me by Sogex as non-collectible dues; and the decision to move for good to the USA was finally taken by Vicky and I.

My work environment that last year in Riyadh was demoralizing to say the least. We worked out of the residence villa of the owners of Sogex, from 10 to 10, six days a week and on occasional Fridays. Bank representatives were continuously coming in and going out; government officials kept the several telephone lines buzzing; friends and foes proposing their help; and Sogex financial staff adamant about producing tons of monthly reports whose values were not worth their weights.

Mentally, I was like a bungee jumper in a whirlwind of worries and concerns. Instead of getting paid, since the Group's funds had dried out, I was paying out of pocket for our living expenses and to meet the Group's obligations towards the Riyadh office staff who were not assigned to any of the projects. By the prevailing employment system in Saudi Arabia at that time, the employer provided the employees with free food, accommodation, and transportation. Senior staff received additionally education allowance to pay the tuition fees for all their children. When Sogex stopped paying and the supply of food and essential services on credit became erratic, I took over out of humanitarian reasons. That also came to an end by May 1989. To add insult to injury, the IOUs and statements of account that Sogex issued to me were continually rejected by the overseers of the recovery plan. The worst part of that turmoil was the unknown future and the lack of a plan to ride it smoothly. I knew one thing for sure: I was heading to the USA and leaving behind Lebanon and the rest of the Middle East.

Philips Camp in Riyadh

During the first year of that dark period, while Vicky and the children were with me in Riyadh, the uncertainty and the frustration were still manageable as hope was not completely lost. Life was dotted with enjoyable moments, funny incidents, and some lucky opportunities. We lived at the Philips compound, thanks to the intervention of my cousin Marwan Junior who was a manager with Philips. Fridays were mostly enjoyed at the pool: relaxing, sunbathing, having lunch, and occasionally swimming. The seventh day of the Saudi week would start with few tricks that I mastered, using marbles or match-sticks, which bewildered Ramy, my nephew who believed that I was a magician. His faith in my magical abilities was reinforced by his witnessing my frequent beating of Amale, his mother, at the Scrabble, when she had him convinced that she was invincible.

Majed's Friends

Majed started his preschool at the Philips kindergarten. He would either walk the short distance from the house, holding Qader's hand, or would drive his pedal red car, with Qader giving him a discreet push every now and then. The trip by the red car came to an end when a wind and sand storm almost lifted him of the ground, giving him the scare of his life at four years of age. When Vicky heard the knocking at the backyard aluminum-glass door, all she could see through the upper glass panel was a group of children, aged maybe 4 to 12, whom she did not know. When she opened the door, Majed, who was the only one completely hidden by the lower aluminum panel, popped out, introduced the group as his Gang and was bringing them over for refreshments.

The Watermelon

The Philips compound, like most of the compounds of international companies, was at the outskirts of the city and was relatively self sufficient. In addition to the offices of Philips, the compound had residential villas for families and senior staff, barracks-type accommodation for the bachelors, a club, sports facilities, swimming pools, supermarket, and a pre-school. The supermarket fulfilled the needs of all housewives except those of Lebanese nationality. The few Lebanese ladies, Vicky and Amale included, preferred to do the shopping at the large malls where more varieties could be found, and at the wholesale food market where vegetables and fruits were sold in boxes, with the exception of watermelons. Saudi Arabia produced varieties of watermelon that compete with Iranian produce when it comes to sweetness and rich red color. A lot of that was sold directly by local farmers off the back of their pickup trucks, usually along the highways or next to residential compounds. The old farmer, who was a landmark when directions were given to locate the Philips compound, stocked his watermelons in heaps close to the compound main gate, under a blazing sun, and waited patiently for buyers: those who know how to pick a good piece; those who pretend to know; and those who admit their ignorance and buy whatever piece is selected for them. I belonged to the last group however, on the one and only one time that I decided to exercise my right to select a watermelon, I went for the largest one believing that the larger it is the better it should be; it was colorless and tasted like distilled water.

Frying Eggs in Istanbul

Vicky was not surprised as she well knows how hopeless I am when it comes to any thing related to food and food preparation. This handicap extends from selection of food ingredients to cooking; from finding the tea bags to locating the teacups; and in fact anything that is related to or has to do with the kitchen. In all fairness to myself I did, after so many years of pumping up my courage, attempt converting a raw egg to a fried egg. It was in the year 2001, when on a mission in Istanbul, and Vicky had traveled to Lebanon for couple of weeks, that I tried to overcome my helplessness around the kitchen and at the same time satisfy the urge to enjoy a lunch of fried eggs. After thorough investigation and few trials and errors, with interval airing of the kitchen to avoid exploding the escaping gas, a shy flame of fire rose from the stove. Selecting the right frying pan and deciding on the amount of butter that should be melted over the fire was relatively easy. Breaking the eggs and ensuring that the yellow and white stuff ends in the pan while the shell fragments head for the garbage can was a task more complicated than I had anticipated. Slamming the first two eggs against each other, like my mother used to do, ended up with the shell in the pan and the egg yolk down my shirtsleeve. Another piece of butter went into the pan, after cleaning the latter from the mixture of burnt butter and eggshells, and a second attempt was made at breaking the eggs, Vicky's style. The third egg, when hit against the edge of the countertop, left the crumbled shell stuck to my hand and the innards cascading towards the vinyl floor. After a new round of cleaning the overcooked butter, the fourth egg was slammed against the sharper edge of the pan; what should have gone in went out and all around, and the shell fragments again landed inside. Finally, I had to resort to sheer brutality: I cut each of the remaining two eggs in half using the bread knife; meticulously picked the fragments out of the soup plate; poured the now clean contents in the frying pan; and a minute later I was devouring a delicious meal of my creation. That meal had become a late afternoon snack and the cause for a one hour of mopping and cleaning to reduce the foul smell to within the acceptable limit set by the World Health Organization and the Environmental Protection Agency.

ILO and IOE

The International Labor Organization, abbreviated and better known as ILO, has been a household name for me since my childhood. Before my teens, it meant Geneva where my father traveled to three or four times a year and brought us back presents. For Amin and I, it was always the special-occasion clothes from Grand Magasin. Lama got her pretty dresses and my mother the very special "Guipure" cloth pieces that were transformed by the Haut Couturier into magnificent evening dresses. Marwan Junior was favored by my father for being the only son of my father's only brother, and so his each-trip regular present was a shirt from Grand Magasin as well. On my graduation day from the elementary school I got a Favre_Leuba watch in addition to the special outfit for the graduation ceremony that I never attended. Amin, who was then 6 years old, also got a same brand watch. He was curious about the writing at the back of the watch which I read for him and explained its meaning: does not break and will still tick when it gets wet. Amin, from that early age, believed and practiced the saying "the proof is in the pudding"; unfortunately, the watch did not hold under the blow of the hammer, limiting Amin's ownership of his first watch to couple of hours.

After her graduation from the Arts & Sciences School of the American University of Beirut, with a major in economics, Lama worked for a short period for the

Chamber of Commerce and Industry in Lebanon and then joined the ILO, where she remained a devoted employee, rising in its ranks, until her retirement. My turn to get involved with the ILO came during our last year with Sogex, while in Saudi Arabia. The Chamber of Industry in Amman, Jordan, held a forum under the auspices of the ILO to develop a plan for the rehabilitation of the Jordanian workers who were returning in large numbers from the GCC countries and in particular from Saudi Arabia. The issue was retraining and redressing their skills to match the needs of the Jordanian local market. The participants represented the labor unions, the Ministry of Labor, Ministry of Industry, the University of Amman, and major employers among the industrialists. As per Lama's advice I submitted a paper on the subject of rehabilitation of labor to the ILO, which was well received, and resulted in being assigned as the chairperson of that forum, on behalf of the ILO. Subsequently, and in addition to the conference that resulted in drafting the international "Safety Manual for the Construction Industry", I participated in a tripartite conference, along with thirty members representing a group of selected governments, employer groups, and labor groups, which was held to study ways and means for strengthening the role of the electrical and mechanical engineering professions in new technologies. My major contribution in these three missions and the important international recognition I received in return prompted me to bring my career as an employee to an end, with the end of the Sogex era, and to launch myself as an independent management consultant.

CHAPTER VI: 1989 to 1993

Time factor and facts of life had a way to dampen my resolve, which had been crystallized during my last days with Sogex, to stay as far as possible from the Middle East geographic area; despite having spent years in a Western environment we remained under the mental and social influence of the Middle East culture and social traditions.

Recruited by Bin Laden Group

When the phone rang at 5 AM on a Saturday morning Vicky and I jumped out of bed, propelled by the reflexes of semi-awareness, and raced towards the source of what we mistook for bad news. We had been in Corona, California, for almost four years, had endured our share of mishaps, and yet our minds had remained set on permanent settlement in the USA. We were by then proud American citizens who strongly believed in the values and aspirations of the American people. The source of very early morning calls was usually Lebanon, due to the time difference, and mostly to convey sad or bad news. This call came from Saudi Arabia, Jeddah in particular, from a person who introduced himself as the human resources director of a major company within the Saudi Bin Laden Group (SBG); the huge corporation that was primarily into mega construction projects. He was well informed of my professional experience and my intricate familiarity with the Saudi construction market, and thus did not have to elaborate on who SBG was. He was calling on behalf of the CEO of that company, who was also a partner with the Bin Laden family, to extend to me an invitation to visit their offices in Jeddah and discuss potential employment in a managerial position. I accepted the invitation, traveled to Saudi Arabia, and had convinced myself that the trip was only to change scenery; after all we were entrenched for good in the USA. Apparently I was kidding myself. The opportunity to return to the Middle East made us realize how nostalgic we were to life that rotated around a large family and a wide circle of intimate friends; how overburdened we were with our problems, with no supporting shoulders to share the load; and how much we needed a change.

On the Move: Houses, Cities and Countries

My intended permanent stay in the USA ended early February of 1993. Vicky waited until the school year was over; took care of packing our household belongings; handled the shipping arrangements; concluded an agreement with a real-estate agent to handle the renting of our house; arranged for shipping my beloved Buick Park Avenue; and moved with the children to Lebanon. When it comes to moving from one house to another or between countries she is unbelievably good and efficient, handling every detail and keeping her calm, while I fret, sweat, and do nothing; we had long agreed that it was better I be away from the house, the city, and the country when the arrangements for moving are in progress. I don't know how far Sinbad traveled and how many places he visited, but I know that to this date we have resided in fifteen cities covering nine countries with a total of 29 houses as primary residence; and my concluded agreement with Vicky held for each related move.

Immigrating to USA

The trip to USA, with the intention to permanently live there, started with an afternoon drive from Monsef to a small town in the Northern part of Lebanon, where we waited till the early morning hours, then continued to Damascus to board a flight to Los Angeles, via Amsterdam. Beirut Airport was closed, and even if it were open, most of the arteries leading to it were clogged, bleeding, or severed. Farewells and goodbyes for a Lebanese family are always difficult, even for a short entertainment trip; imagine the emotional outbursts in the face of immigration to America at a time when the roads out of Lebanon were extremely dangerous, and the possibility of return was totally unknown. Despite the advanced technologies that had transformed the World into a single global entity, the USA was for the Lebanese elderly generation the "Other World". Yet, Vicky's parents understood that we had to leave Lebanon, if not the Middle East region altogether.

The two previous years were as hard as could be on all of us. Vicky and the children spent more than a year in a 600 sq. ft. apartment in Monsef while waiting for the old Ghosn villa, which I had bought earlier, to be vacated by its previous owner. I spent the time traveling between Saudi Arabia and Lebanon in my fruitless efforts to collect my money entitlements. The children went to Monsef School where they received, in addition to the normal curriculum, intensive Arabic. We knew that was their last chance to acquire the basics of their native language since, despite the special lessons they had been taking while roaming around Europe and the Middle East, their speaking and reading skills were less than elementary. Vicky's parents took advantage of a quiet break in the ongoing turmoil and moved for good from Beirut to Monsef, to reside in our newly acquired villa. What they lacked in house conveniences when they moved in, which were many considering that the villa was built in phases and in sections over more than one hundred years, they made it up by furnishing it well, supplying a large capacity standby generator, and installing a temporary gas-fired heating system. Rehabilitating the old villa started at a slow pace and did not pick up speed until after Mira and Majed had graduated from their respective universities; Vicky's parents had passed away by then and, sadly, did not share with us the joy of transforming an old building into a warm and modern residence.

Los Angeles

Self-employment, the third and last phase of the business category of my life, started when the plane landed at LAX airport in Los Angeles, California, in August of 1989 on a day and date that I cannot recall. Ghassan and his brother Khaled, whose mother is Vicky's first cousin, met us at the airport and drove us to the Quality Inn motel in Anaheim, where we spent two weeks in the same rooms that we had been staying in, on our previous yearly visits to the USA. It is also the same motel wherein Samir, my closest friend from the Group Seven, had introduced us to Debbie, whom he married a year later. Samir had first met Debbie in Riyadh while she was visiting her brother, a colleague of Samir's, and fell in love with her and with the concept of having a girlfriend. Now he lives and works in Qatar and regularly visits his home in England, near Manchester, where Debbie watches over Alex and Sara. He was here with me in Kuwait over the weekend, on a short visit, to catch up on the past four years' news since we last met. While he went for his Doctorate degree I went to work with Orient; now I am anxiously waiting for Majed to complete his fourth year of work experience and start his MBA while Samir is still waiting for Sara to finish her remaining three years of high school.

Corona

First priority was to move out of the motel and settle in Corona, Riverside County, where we had already decided to establish residence for several reasons: the public education system was among the best in Southern California; the place was safe and environmentally superior to most of the nearby cities; the community was friendly and neighborly; housing prices were affordable; Corona was among the fastest growing cities nation wide; and Ghassan and his family lived there. I opened an account in a local bank in Corona, picked my temporary checkbook, and went with Ghassan to buy a car. Ghassan recommended I buy it on credit against a loan in order to establish a credit rating, which is important to conclude most of the financial transactions in California. Apparently cash was not the Master and the King like it was in the Arab World and even in Paris. The car dealer refused to give me a loan since I had no credit rating, and I had no credit rating because I never had a loan before; the story of which came first, the egg or the chicken. So we shopped for a loan and got one at high rates. The car dealer would not accept cash for the down payment and would not accept my check since it did not have a permanent address on it; I had no address. I told Vicky: I get the message; the American way is not for us so let's go back. Ghassan used his checkbook to comply and the sale finally proceeded. I was offered a "used car repair insurance policy", which was available at the time of purchase only, and I declined it to be penny wise and pound foolish. I drove the car straight from the dealer to a car mechanic for a thorough checkup that lasted two days. It had a dormant A/C compressor problem and a potential automatic transmission slippage that required more than my skills to detect them. I went back to the Buick dealer who absolved himself from any responsibility and blamed me for not buying the optional repair policy. It is my good luck to run into an "experienced" Lebanese when faced with car related problems; he overheard the argument from his spare parts sale counter; could tell from the "language" I used that I was also of Lebanese origin; and suggested that I still had one more day to return the car under the State return policy. I did, then re-bought it back at the same price, with the appropriate insurance policy, and drove away an hour later.

That same insurance policy proved the common claim that the American System treated the consumer, in addition to the taxpayer, as a king. We were driving back from Reno, Nevada, and had reached the outskirts of LA after seven long hours of enduring three exited and tired children, when the potential automatic transmission problem became a reality. I barely managed to make it to Corona and reported the problem to the insurance company the following morning. The manager of the car dealership I was referred to asked me to bring in the car 3 days later, since his transmission specialist was on leave; I did. The following day the manager called me and gave me the bad news first: the transmission was beyond repair and had to be replaced. Second, he gave me the worse news: the insurance company advised that the policy had expired the day before and so the repair cost had to be borne by me. The fact that the problem had occurred and was reported prior to expiry of the policy, and that the delay was caused by the Buick dealer, were inconsequential to both parties. The "Better Business Bureau" and Buick's customer relations office had something else to say: the dealer furnished the transmission unit, the insurance company paid for all labor costs, General Motors sent me a letter of apologies, and I ended up with a new transmission. The Lebanese spare parts sales person who had shared with me his knowledge of how California return policy worked was the recipient of an expensive cigar box.

Concluding the purchase of the house on West Citron Street took a bit longer than the car to approve the mortgage. It was the model home for that track, on a corner lot, with well manicured front and back yards, and tall Eucalyptus trees lining the backyard wooden fence. It was on a windy night that we woke up to a thundering sound to discover that one of these trees had fallen down, barely missing the house, damaging the fence, and blocking most of the road at the front of the house. During the three months waiting time we stayed in a rented apartment in a decent and secure compound, but with an unsuitable address. It was then and a bit too late that we found out the school vs. address relationship: the home address decides what public school the children are. allowed to go to; and it was not the school of our choice. As fate has it, the way out was advised by a Lebanese living in Corona: appoint a guardian for the children whose address corresponds to the school of your choice; we did and the matter was solved.

Schools in Corona

During the five years that we lived in the States, all the time in the same house in Corona, Majed finished his elementary schooling; Mira completed her elementary and junior high; Marwan completed junior and most of senior high education. Apparently the prevailing belief that constantly moving negatively impacts children's education was not true in our case. Maybe because we maintained for them the same International-American educational system, but definitely because the home environment was consistent: caring, supportive, guiding, not imposing, and inducing to developing the children's character and personality. I still keep their ribbons, cups, awards, and achievement certificates that they got for academic achievement, sports, and perfect attendance. Each of them received at least one President's (of the USA) Award; the attention and encouragement that elementary and high school children receive is unparalleled, at least within the Corona school district. The best part was the advanced placement and the gifted students program: students stayed within their age group but could academically advance at their own pace and are not hindered by slower learners.

The house was well situated, within walking distance to the Junior and Senior High Schools, and couple of minutes by car to the elementary school. It was called Foothill Elementary because when built it was on a hill, among orange groves, overlooking the city to its south. When we visited Corona in the year 2004 on a nostalgic tour, the orange groves had been replaced by a forest of housing developments. The park next to our house was still a park, and that is where the children spent a good part of their off-school time. I believe our long service in the Arab Gulf states left us park-deprived, and that is why greenery, which was abundantly available in Lebanon when I grew up, attracted us wherever we took our children. Whenever weather permitted, and that was frequent in Southern California, Ghassan and I took our families on picnics where the grownups relaxed while the children rode bicycles, patted domestic animals, played football, and let out steam. Yara is about six years younger than Majed and she took her first unaided steps in our backyard, which I witnessed to the envy of Ghassan. Suhail was born almost a year after we moved to the USA and I took it upon myself to spoil him rotten. He is now in his mid teens and should probably be listed in the Guinness Book of Records, thanks to me, as the youngest person ever to have puffed a cigar, drank whisky, and be kicked off the gambling floor in a casino when he was not yet one year old. He is now seventeen years old; never has repeated and I believe will be a long time, if ever, before he repeats any of those one time experiences. I think I did well by him like my grandfather did for me when I was 5 years old.

Smoking my First Cigarette

It was during the summer of the year Amin was born in. My father was on a business trip in America; my uncle Hassan was not yet married; our house in Monsef was undergoing a complete renovation; my grandparents, on my father's side were living with us; and we were all staying at my grandmother's house, my mother's mother, in Monsef. On one afternoon I approached my grandfather while he was rolling a cigarette using local tobacco and asked him for one as well. He readily obliged and lit it for me at the same moment that my uncle walked through the front door. He was stunned, furious, and would not listen to my grandfather when he told him that he knew what he was doing. I ran away while inhaling deeply and quickly with the cigarette dangling from my lips, perfectly imitating my grandfather. My short, young legs did not carry me far enough before I collapsed, hardly breathing, and instantly developing a high fever that left me bed ridden for few days. My second smoke was the year I graduated from AUB. Now I smoke cigars, maybe 3 or 4 a week, an interest I acquired while living in Paris, which I am not addicted to, and only enjoy after a good meal, mainly a dinner with friends.

FANA

Our house in Corona was also about five minutes' drive from our offices. Ghassan owned the one-story building which was the offices of the land development company he co-owned with a local partner. The wife of this partner was an executive with the California Board of Education, and she was the one to recommend the Corona school system. Khaled assisted his brother Ghassan, and their cousin Imad worked for them. Ghassan, Khaled and I registered a company which we named FANA, a name composed of FA from Farah and NA from Nasr, our respective family names. FANA's first business venture was buying a Pizza franchise, Double Deal Pizza, and leasing and furnishing the takeout/delivery place. A prerequisite to buy the franchise was that the owners/managers, Ghassan and I, had to go through a two-month training program culminating in written and practical tests which we both passed. I had befriended the director of the Franchise Company during the training course to solicit his sympathy for my cooking abilities, a skill that I could never come close to acquiring and performed as badly as frying the eggs in Istanbul.

Double Deal Pizza

We came to know how to manage the Pizza Store; how to hire and train staff who were mostly college students with the exception of the store manager; how to run, clean, and maintain the automatic oven and the cold storage area and refrigerated enclosure; what ingredients to order and how to estimate quantities; how to manage the books; how to prep the pizza; and most importantly how to treat and handle the dough. Anything that had to do with the food part was transferred to Khaled who, having lived on his own during his student days in Arizona, was innovative and good at it. I kept the management part for myself, or at least I did that until Marwan was diagnosed as having a brain tumor.

GRF and DDP

The first two months of operation we received the award for the best performing novice place among the fifty or so franchisees that were established within that year. Our pocket pizza was the talk of the town; our prices were the most competitive; our customer follow-up was highly appreciated and was reflected in the tips the delivery staff received. On weekends, we all pitched in to keep delivery time within 15 minutes and save on giving free pizzas. My role was restricted to delivery and I enjoyed the tip as much as the next person. By the end of our first year of operation our cash flow turned from positive to negative. Nation wide recession, which had started a year earlier, caught up with California; The Golden State usually lags behind the other States when recession hits or when recovery starts; and our starting of a new business in Southern California was ill-timed. To make it worse, Pizza Hut, the Company store and the franchise, which lost a good percentage of their market share to us, retaliated with a massive and aggressive advertisement campaign that we could not even dream of matching. The third factor that expedited our fall, with no hope of slowing it down or reversing it, was when Marwan got sick and I started ignoring the business all together. Three years later, with most of my reserve capital down the drain, we closed shop. The only business that kept me partially occupied was GRF, a company that George, Vicky's brother, and I had started earlier to handle import of foodstuff from Lebanon to a wholesaler in Pasadena. We established GRF after the invasion of Kuwait by Iraq to avail a legal status in the USA for George and his family in case the war spilled over the rest of the Middle East. The impact of the invasion was localized, and as a result George and his family never relocated to America.

Marwan's Illness

Marwan was on the basketball team of the Corona High School but had stayed at the bench for few consecutive games as a result of a headache that started as soon as he ran, and then subsided when he sat down. His doctor recommended we see a specialist after ear and sinuses tests showed nothing. We called a neurosurgeon friend of ours, a Lebanese doctor with UCLA medical department whom we had met through Amin, who had done his internship in Iowa at the time that Amin was specializing in Oculoplastic Surgery. It was on a Sunday morning and he asked us to come over to his house for preliminary check. On the way there I got pulled off by a traffic officer for tailgating and was issued a ticket; I did not know then what tailgating meant and did not care to ask. That was my second traffic violation in almost 24 years of driving. The first violation was in 1967, the day I picked my first driving license, when I was stopped by a Lebanese patrolman for overtaking another car while on the approach road to a bridge. He accepted my argument but insisted on the ticket so as never to be careless and always get home safely for the sake of my mother.

Dr. Comair had Marwan walk in a straight line with his eyes closed, touch his nose with alternating hands, and do some movements. He hid well his worry when explaining his recommendation for an immediate MRI (magnetic resonance imaging), to rule out the bad possibilities, which he arranged for at UCLA medical center. Few hours later Marwan was admitted for an urgent surgery to remove a tumor diagnosed as Medulloblastoma. From then on our emotional life was marred by a wound that time could never fully heal. His operation, which took eleven hours, was characterized as successful. We recruited friends and family members in the medical profession to assist in seeking consultations from as many specialists as possible; the unanimous decision was a follow up course of chemotherapy and treatment by radiology, all within a newly developed protocol that Dr. Bedros advocated. Once a week I would pick Marwan up from school at the end of the school day, rush to Loma Linda Medical Center for his one hour combined treatment which was preceded by blood and other tests, and then get back home in time for a late lunch and resumption of normal life; not once did he complain during the one full year of treatment. On the sixth anniversary of his operation, two years after we had relocated to Lebanon, we went to Loma Linda for Marwan's supposedly final checkup; the Doctors declared him officially cured and we celebrated Christmas of 1996 as close to God as humans can feel despite the fact that we were surrounded by the vice of dice in Las Vegas.

CHAPTER VII: 1993 to 1998

My assignment with Bin Laden Group was to spearhead the transformation of their electrical and mechanical construction company into an industrial unit which would undertake major water, power and oil projects. After a short while I was convinced that such a transformation was not a priority, as I was led to believe, and that the company had needed me for my curriculum vitae, for the director position, in their submission of a proposal for a large power and water desalination project. The project award went to another company and so my duties were quickly redirected to the Group's efforts in the acquisition of other companies. A team of two, the Director of Finance and I, went after a large fabrication company, owned by an Italian family, located in the Eastern Province of Saudi Arabia, specialized in making vessels for the oil industry,. The negotiations took the best part of 1993 to iron out all the financial, technical, and legal details to the satisfaction of the Group; however, taking the final decision to conclude the takeover hit a stumbling block that remains unknown to me.

By the end of that year my family had been in Lebanon for almost six months; my disappointment in not seeing the fruit of my efforts had reached a crescendo; rumors about financial difficulties of major government contractors in Saudi Arabia was drumming loud; and the belief that Lebanon was heading towards an unparalleled boom gave me enough incentive to say goodbye to Jeddah and hallo Beirut. What gave me the final push towards Lebanon was the offer I had received, while still on assignment with Bin Laden Group, for a two-year contract to manage the construction of Beirut Sports City. On one of my short visits to Lebanon Moufeed had asked me to help a friend of his, also a minor partner in

Moufeed's business, in reviewing and commenting on the draft contract agreement his friend was about to sign with the English Prime Contractor for the construction of the Sports City. I met Elias, a likable person, and promised to review the documents over the weekend. On Monday, Elias went over the contractual notations I had highlighted, was overwhelmed, and asked me if I would attend a meeting that afternoon with his counterpart from the Prime Contractor. I lead the discussion and negotiation of most of the clauses; the Prime Contractor adopted most of my recommended changes; Elias was happy; and I was offered the project director's position.

Beirut Sports City

Beirut Sports City had been a major achievement when built in the nineteen fifties and a prime casualty during the subsequent Lebanese war. It was said to be a stronghold for the Palestinian fighters, with underground passages leading to adjacent camps and large storage depots for arms and ammunition; a tempting target for any entity that objected to the increase of Palestinian power in Lebanon. The post war Government decided to make the rebuilding of the Sports City a symbol of the rebirth of modern Lebanon, and the first project to be rebuilt with international standards. The Main Contractor, who was acting as a Construction Manager, had hired a sub-contractor to execute the works. The sub-contracted company, after three months into the contract, had not started work on site and was duly terminated, leaving all concerned parties hanging in midair. The project was supposed to be on a fast track with a deadline to be met; and the Pan Arab Games were to be held in Lebanon as a recognition of the return of peace, with full financing by certain Arab States. The financing of the reconstruction did not fully materialize resulting in down-scaling the scope of work and restricting it to the major Stadium and the Multi-Purpose Hall.

Back-to-Back Sub-Contract

The new Subcontractor of the civil works for the Sports City, a firm co-owned by Elias, had entered into a back-to-back sub-contract agreement with the Main Contractor. All contractual responsibilities of the Main Contract relative to civil works were passed to the Subcontractor including coordination with the Client. The financial implication of back-to-back is that the Subcontractor issued Performance Bank Guarantees in favor of the Main Contractor for the same amounts that the latter had issued to the Client, and the Subcontractor gets paid for the work it executed only after the Main Contractor gets paid by the Client against the same work. Accordingly, if the Main Contractor fails in his coordination and overall management, or his only other minor subcontractor for the electro-mechanical works is in delay, the Client might stop payments or, in a worse case scenario, he might call in the bank guarantees; in both cases the Subcontractor is the party to be hit hard by such actions. That clause would not be waived by the Main Contractor during the negotiation of the Agreement, and despite my warning Elias took the risk and accepted it.

Elias had bargained on his good relationship with the government official who was instrumental in bringing him as the subcontractor for such a prestigious project. Furthermore, he was over zealous in his quest for recognition in the construction industry as a top class performer, and maybe in the political arena as a respected player. The Subcontractor's establishment in Lebanon was founded on the construction company that Elias had worked for, over many years, in the rest of the Arab World. My mission was to set up the organization of the project and its facilities to transcend the project demands and constitute new elements in the company; the Sports City project was to be the launching pad for potential involvement in the highly talked about construction boom of Lebanon.

From day one we were in a race against time, lacking almost all the necessary tools, and on a track with bumps at every corner. While we were building our site offices my car was a filing cabinet cum working table on wheels. Site excavation started on 24-hour basis, and was only stopped for the short time that it took the army to defuse the unexploded shells and ammunition that were frequently dug out. At the project lay-down area the rebar plant, concrete batching plant, pre-stressed concrete factory, pre-cast factories, workshops, garages, warehouses, and open storage yards were coming up all at the same time that the workers' facilities were being built; for a fast track project the labor force had to sleep, eat, and drink on site and be willing to work 7 days a week and sometimes two shifts per day.

Project Components

Building up an organization to handle such a construction project had to start with identifying and defining the elements involved. These elements fell under the following categories: Engineering & Design, Planning & Scheduling, Material & Procurement, Office Administration & Accounting, Quality Control and Safety Procedures, Quantity Surveying and Cost Control, Offices & Accommodation, Lay-down Areas and Productions Facilities, Manpower, Equipment, and Financing & Bank Facilities.

Engineering & Design

The Sports City Project is mainly a building construction project where the architectural design and the engineering were contracted to an independent architecture and engineering (A&E) firm. However, *site engineering is essential and it mainly involves preparation of shop drawings to translate an engineering drawing into as many detailed sketches as needed, which would then be used by the artisans on site to build from. By this way an element can be broken up into components that relate to different specialties for ease of understanding. Site engineering is also important to develop design alternatives in order to simplify and speed up execution, reduce cost, or deliver improved product without jeopardizing the intended quality of the finished works. Cost savings to a contractor by speeding up execution, simplifying the construction, and handling the design of additional works, more than pays for a site engineering unit.* Also, as required by the conditions of contract of the International Federation of Consulting Engineers (FIDIC), which were the conditions applying to the Sports City Project; we were responsible for identifying design or engineering deficiencies in order for the A&E firm to rectify them. The argument is: *if a contractor had no engineering responsibility then the Client would have required the Contractor's staff to be lawyers or doctors rather than professional engineers. It makes sense.*

Major design changes were initiated by Nabil, the Site Engineering Manager. Concrete seats, fifty thousand of them, had to be redesigned to permit on-site precasting for speeding up the execution. The Stadium cover design had to be modified to suit in-country available lifting equipment. The round metal roof of the Multi-Purpose Hall had to be redesigned as a space frame by a US engineer, who claimed to have designed the legs of the lunar module that played a part in the conquests of the moon. After his visit to Lebanon, an act that was not condoned by the US Government, he wrote his Senator describing the peaceful and safe life in Beirut.

Planning & Scheduling

How often are we asked what our plans are for the weekend or for the holiday? Why do we plan; when do we plan; how detailed should our planning be; these are questions that could be better answered when we decide on the definition of "planning". To me, *planning is defining a logical path to reach a pre-set and attainable goal. We plan in order to avoid disruptions, distractions, and possible failure prior to achieving our goal. Planning is a must if we have constraints along this path, such as time and cost. The level of detailing should optimize the number of activities and their complexity. The activities or actions should remain manageable, controllable, continuous, and most importantly one-directional with a focus on the required goal.* By telling a child: time to go to bed; put on your pajamas; and brush your teeth, a goal is set and the activities are continuous, comprehensible, and achievable. A non-argumentative child would be in bed in ten minutes. Telling the same child that it is already past his bedtime; to get off the sofa; climb the stairs; go to his room; undress; wash up; put on the clean pajamas in the upper left drawer of the dresser; brush his teeth; is an example of very bad planning and overkill. The goal was lost sight of and you would need a re-run of what to do. On the other hand, telling a ten-year old boy to brush his teeth and go to bed immediately if he wants to become a successful medical doctor has set a goal with a very poor plan.

The plan for executing the Sports City was then transformed into a schedule following the same procedures that were applied for the Yanbu project; however, the computer program utilized was more recent software, user friendly, and offered better monitoring and control of the program of works.

Telling the story of planning and scheduling is boring. Preparing the plan is tedious. Updating the program periodically to reflect the actual start and finish dates of each and every activity requires patience, perseverance, and meticulous work. However, a project badly planned or its schedule inadequately updated is destined for failure.

Planning is logic; scheduling is knowledge; and compliance with both is self-discipline. Engineering is a good conduit for developing logic and knowledge. Discipline is a personal characteristic trait that most likely has its own gene somewhere along the DNA chain. Unless it is practiced, discipline loses its glitter and eventually fades away. However, it should be practiced in moderation with allowance for spontaneous actions; at least that is what Vicky keeps telling me in order to avoid overstress and its unhealthy consequences. She knows well my subconscious tendency to pre-plan everything; I even planned to marry her long before I knew her. I bought my first portable computer, the famous Kaypro, with its special CPM scheduling software to have a planning tool at my side wherever I go. My first executive position was: Vice President Planning and Controls. My current job is a management consultant for the Ministry of Planning as well as the Supreme Council for Planning and Development for the State of Kuwait. Dwelling so far on the planning issue was a spontaneous act triggered by that persisting gene.

Material & Procurement

Engineers and designers must not restrict contractors to a specific brand for any material item unless absolutely necessary. The contract documents should state material specifications in accordance with the requirements of the project. If guidance is essential, due to limited suppliers or complex details of the item or other compelling reasons, the engineer could mention a brand name and add "or equal". This would prevent design engineers from favoring one manufacturer over another for personal gain, provides equal opportunity to all suppliers, and gives contractors a wider choice to seek competitive prices and better quality. Adding a fourth pump in a pump station that should be synchronized with the other existing pumps is a case that could allow the specifying of a brand, type, and even model.

Every element or component or stand alone item that ends up a physical part of the final project is defined as material in the contract documents. It could be the double leaf wooden door listed under the schedule of doors on a drawing; the sand with its specified grading to be used in the concrete mix; the pre-cast concrete beams with defined physical characteristics and manufacturing procedures; or the Aluminum roof space-frame structure for the Multi-Purpose Hall that was designed by the American lunar-module-legs engineer; manufactured by a specialized shop in Florida; shipped by a Belgian freight agent on a Greek registered vessel; and erected by a Lebanese engineer on a sabbatical leave from his employer in Abu Dhabi. *All these materials are identified in the contract Bills of Quantities (BOQ), the document that lists all project items, their quantities and prices which, when added up, form the total contract price. Material items are listed as supply and install.* Systems, like air-conditioning, include testing and commissioning. The BOQ could also have items that are services only, with no material included, such as mobilization to and demobilization off the project site.

Sale of Excavated Sand

Excavated sand and soil was the only material item in the Bill of Quantities of the Sports City project that we were paid to get rid of rather than supply it. The contract required that the excavated material be transported and stored off-site in areas designated by the Client; a straight forward instruction if it were not for a side agreement that had been reached with the authorities giving ownership of this sand to the Main Contractor. Buried unexploded shells were detonated; large quantities of discovered ammunition were turned over to the army; the remains of animal parts, in the covered dump area where the slaughterhouse used to exist before Israeli war planes flattened the place, were hauled to incinerators; few thousand cubic meters of sand were heaped on site for re-use as backfill material; and the remaining hundreds of thousands of cubic meters of excavated sand were transported off-site under the control of Asa'd, the excavation sub-contractor.

It was 7 AM on a Friday when an internal security jeep came to a stop in front of my office on site, and two uniformed officers stepped down and invited me, ever so politely, to go with them to the internal security headquarters for a matter that they were not at liberty to disclose. I could not refuse the invitation however the officers were kind enough to accept an alternate means of transport; the representative of the Prime Minister, who was assigned to monitor the daily progress of work on site, had just arrived and in an authoritative voice advised that he would give me a ride. After five hours of waiting and waiting in a sparsely furnished office, visited only by an elderly man who kept me alert with his dark strong and reasonably tasty Turkish coffee, and armed with my cellular phone that kept me in contact with the outside forces that were relentlessly trying to solve the mystery of my civilized arrest, the story started unraveling. In the early hours of that day, a tipper truck over loaded with sand was stopped at a police checkpoint. The driver, when questioned, advised that the sand was for the Sports City and that he was delivering it to someone who had bought a large quantity of this sand from a nearby stockpile. He knew only my name, which he gave freely, and advised with all confidence that Walid Nasr is the only boss and nothing happens without his approval. He had somehow heard my name in association with the Sorts City, and sincerely believed that throwing it around could get him out of whatever trouble he was in. The driver not being my employee; the sand not belonging to my employer; and the sale transaction having nothing whatsoever to do with me did not stop the over-zealous policeman from deducing that the sand was stolen and that I was party to it. At around 2 PM the lawyer of the Main Contractor's local partner called me from London to advise that that issue had been cleared and I should be released immediately. The "immediately" took three hours for the General Prosecutor to agree to sign the release papers, since he had already started enjoying his weekend much earlier on that day, and had a policy of not being disturbed during his rest hours.

To this date the facts related to the sale of sand are not clear to me; rumors about the right of ownership for excavated sand from government owned property and transfer of this ownership were abundant and remain a subject for social gossip; the list of beneficiaries from that sale increases or decreases on speculation only; and the profit made is a guesstimate but should have been a lot since the price of cubic meter of good quality sand had more than tripled during that year, driven by the government control over sand and stone quarries. The lawyer who called from London became a good acquaintance, and few years later, as the minister of communication, authorized a telephone line for my apartment in Beirut when lines were unavailable, physically, administratively, and politically.

Quality Assurance & Quality Control

Quality assurance and quality control, or better known as QA/QC, developed fast in the manufacturing industry, where chain production and automation have lead to fast repetitive work and to a decrease in human intervention during the manufacturing process, which necessitated the development of control procedures to prevent faulty products from reaching the consumer. The construction industry, whether for large scale housing and commercial developments or for industrial facilities like water desalination and power plants, is becoming more and more complex, highly expensive, and continuously suffering from worldwide shortage of skilled labor. *The responsibility for assuring a good quality final product, equivalent to what the owner paid for, has shifted from the individual worker to a QA/QC program managed by specialists whose function is to monitor the quality of work, guided by the procedures of the program.* International quality standards have been developed, and recognized agencies have been authorized to review management systems and operating procedures of all types of companies in order to certify them as compliers with the relevant international standards. In the current highly competitive markets and the increasing ease of transporting technologies and labor among nations, regulatory and control procedures for assuring quality work and quality product are a must.

Health & Safety Procedures

Health and safety program is another requirement that is becoming a must for all projects executed under strict quality standards. Here, *the procedures and regulations are intended for assuring the health, safety, and survival of the people associated with the project. It extends as well to the protection of the works from damage; the safety of equipment used for construction; and the safety and security of the means of access on site.* Preparation of daily safety reports was mandatory at the Sports City project, and the international Main Contractor published a monthly bulletin, with curves and graphs, reflecting the number and type of incidents on site. Wounds from stepping on nails had the lion's share of injuries, keeping the two male nurses at the on-site first-aid facility fully busy. At the beginning of the project I was adamant about adhering to the safety regulations of the "Safety Manual for the Construction Industry" that I had participated in writing at the ILO. Every worker on site was issued a safety boot the day he was employed, and only after attending a long orientation session about safety and safety procedures he was allowed to go on site. Still, the prevailing footwear was a tennis shoes or a rubber slipper, and both were inviting to nail injuries. I kept insisting on providing the workers with replacement safety shoes, which were expensive, believing their claims that the old ones were lost or stolen or damaged. Revelation came on a Saturday afternoon when I witnessed, firsthand, the exodus of large groups of workers who were heading home to enjoy a three-day religious holiday with their families; they were in their rubber slippers with brand-new pairs of safety-shoes, shoelaces tied, and casually thrown over their shoulders. Nails or no nails, no more safety-shoes.

Accommodation & Lay-Down Area

Contractors are usually allowed a mobilization period at the beginning of the project to provide the necessary resources and facilities for the start of work. Large projects like the Sports City requires a three-month period; we had none. Building offices, housing for the laborers, pre-cast concrete manufacturing plants, concrete batching plant, warehouses, storage areas, garage, re-bar cutting and bending plant, and stores went all in parallel and at the same time that equipment were purchased or rented and manpower recruited. Seven days a week and 24 hours a day were not enough to meet the construction schedule and the fixed completion date that was tied to the Arab Games, the vote of confidence from the Arab States that Lebanon was out of its civil war and on a fast track for recovery. The project was of such political significance that the Prime Minister of Lebanon used to visit the site every Saturday afternoon; inquires about the weekly progress; asks technical questions that only a highly experienced contractor would think of; comments on production rates of re-bar and concrete; and repeats his instruction to call him day or night if faced with any problems that could impede progress.

Elias had insisted at the beginning to have his office on the roof of the broken down Multipurpose Hall, which was destined for demolition and re-construction, despite the fact that there was no still-standing access to the roof. He wanted to have an unobstructed view of the whole site. Eventually, when the site was excavated deep enough to the final level for the start of work, his naked eyes failed to register all the details as they were not biologically created to see clearly from that distance. Reluctantly, he stepped down from his loft and settled for a porta-cabin strategically placed at the edge of the excavation.

The up-in-the-sky office of Elias was the least of the mobilization problems. The Syrian army, which originally came to Lebanon as part of the Arab Deterrent Forces at the start of the Lebanese civil war, had a military post, manned by less than ten soldiers, on site where the offices had to be built. Relocating their post to a vacant lot outside the perimeter of the site, fifty meters away, took almost two months of discussions, negotiations, and political intervention.

Syrian Intelligence Officer

Building accommodation facilities for more than one thousand laborers on site was probably a first in recent times in the construction industry in Lebanon. The challenge was keeping peace and order among a majority of Syrian workers, a minority of Lebanese, and few Egyptians. Credit goes to the Syrian intelligence officer who was in charge of security in a neighborhood suburb of Beirut. I was in my office; where else would I be, when my secretary announced a pleasant looking gentleman, in civilian clothes, who wanted to meet me. My motto was to always have an open door policy and to spare time for any caller, especially that I had trained my eyes to read, my ears to listen, and my hand to write simultaneously, independently, and with equal focus; so I welcomed him. He introduced himself as the head of intelligence for a large geographic area east and south of Beirut, and wanted to make sure that all is under control among the laborers. From his accent I assumed he was Lebanese from the North of Lebanon. He was a Syrian who had been posted in Lebanon for more than 15 years; married to a Lebanese; his children were born and raised in the suburbs of Beirut in a house that he owned; and that he had not been in Syria for more than three years. "What is the maximum duration you are allowed to serve in Lebanon before you are called back", I asked. His matter-of-fact answer was that originally it was a two-year period, then minimum two-year service back home before another foreign assignment is allowed. However, the Syrian officers were deeply involved in the routine life of the Lebanese people; the logical conclusion was to keep those who have adapted well in Lebanon until they were due for final release from active service. The truth in what he said hurt my pride, but his honesty and sincerity dampened my frustration. He offered to send me an intelligence officer, whom he trusted and vouched for his discipline, to be hired as a camp-boss to ensure calmness and harmony among the laborers. I accepted, and the clean bill of proper behavior at the camp and on site was a proof of the Officer's good intentions and my good judgment.

Manpower

Production, efficiency, and quality, all boil down to the human element. Management sets policies, provides guidance, monitors performance, avails the tools, and hopes that the evaluation and selection criteria for hiring appropriate performers are effective. Bechtel hired me after a long and pleasant discussion about politics, arts, and literature; I was not even aware that it was an interview session. I hired Ramzi after he demonstrated his ability to prepare a program of works in the form of a network and a computer generated schedule. Jaber recruited for the Sports City project any laborer or artisan who walked through the gate of the boundary fence and claimed to know anything about civil construction. We had to reach a labor force of almost 2000 within a very short period, in order to catch up with an inherited delay of almost three months when we signed the contract with the Main Contractor. The post-war construction boom had started in Lebanon; skilled artisans were a scarcity; and beggars could not be choosers. *Re-assignment and re-re-assignment of workers to different tasks and a continuous on-site training were the means used for "selection of the fittest – most suitable for a task-".* The level of hiring and firing remained at a crescendo with a gradual increase of the first and decrease of the latter until we reached the plateau of sustainable production, efficiency and quality of workmanship. The on-site accommodation, overtime work, and a slightly better pay than the market eventually attracted the higher skilled people; we did not steal workers from other contractors, at the same time we did not want to know where they came from.

The secret to getting the best out of a person is the fact that every person has something positive to offer. The successful manager, whether he is a foreman, supervisor, manager, or director, is the one who could bring out the best in others in the shortest possible time. Human resources management has become a science and a necessity for all industries, from manufacturing to services to construction. *Recruitment is the selection process; training is for development of skills; procedures are for guidance to better performance; and work environment is for improving productivity. Personal treatment is what binds all these together to ensure sustainable growth*; it was what induced the manpower on the Sports City project to work unlimited hours and non-interrupted days, and to achieve international levels of quality and production. Joseph, the houseboy in Jeddah, became a welder; Mathieu, another houseboy in Yanbu, became an assistant administration manager; and Walid, the electrician-helper at the Sports City, became one of the best electrical contractors in Lebanon.

For every rule there is an exception. Joujou, the electrician at Sogex: the more benefits and chances I gave him, the more demanding and less productive he became. Fouad, the manager of the engineering office in Kuwait: Guiding and supporting him to win the managerial position, against other qualified competitors, improved his backstabbing skills, and completely eroded his trustworthiness.

Rony

Rony joined the company within SBG, in Jeddah, almost the same time that I did. He was the Human Resources Manager with a mandate from the CEO to set policies and procedures for hiring, training, managing, and firing of employees. The firing policy was the simplest and easiest of all: at the rate that senior employees resigned would have made such a policy redundant. Rony soon came to the realization that *policies are meaningless if not enforced and implemented*; something he had no control over. His eagerness to contribute towards the realization of the CEO's vision equaled mine and his frustration due to lack of implementation surpassed mine. So, we decided to join forces and subject the senior employees to management techniques through weekly seminars. When it came to managing the human element, Rony and I complemented each other: he had in-depth theoretical knowledge and I provided the extensive hands-on experience. We always delivered the seminars jointly, in an informal and relaxed atmosphere, and encouraged the attendees to participate in presenting, discussing, and debating new ideas. The participants gradually grew from few company employees to a large group of senior staff from other companies as well.

Misinterpretation of management techniques was addressed in most of the sessions, related to the subject under discussion, by giving true examples narrated in a form of a story to give them prominent places in the memory filing cabinet. On the subject of meticulous work: The director came across one of his managers in the hallway; he started shouting at him and accusing him of lousy work, bad performance, and sheer stupidity; and walked away without being specific or giving the manager the chance to think, ask, or react. As the director explained later, the manager had done nothing wrong. The harsh encounter would make the manager go back to his office; re-check everything he did for the past several days to see where he went wrong; correct his mistakes if any; and decide to double and redouble his review of future work to make sure that no mistakes are made. Moreover, the open space offices with low partitions carried the director's monologue to all other employees inflicting on them the same reaction as that of the suffering manager. A meticulous person has been created? Maybe so, but definitely another ex-employee is added to the list of quitters.

On the subject of authority and responsibility: A director wanted to restructure his company's organization to encourage decision making, quicker actions, and more efficient work. The concept of decentralization of management by giving the departments' managers control over their own functions, defining the linking relationships among the various departments, and clarifying the lines of communication with the board of directors, was presented to the managers, in a general meeting, graphically in the form of satellites. The managers started arguing about the linking relationships between the satellites, each wishing to have more authority lines going out of his satellite than responsibility lines coming in. The director intervened and delivered his ruling as to the concept of satellites: He was the transmitting satellite, like Arabsat or Nilesat satellites that transmit TV programs, and every department manager was a television set that received instructions from him and only him. Obviously, restructuring efforts were put aside and never attempted again.

Financing & Banking Facilities

On my tenth birthday my father opened a savings account for me at the British Bank of Lebanon, located in downtown Beirut. Today, after half a century, I maintain that first account but with minor changes: The bank's name is now HSBC; the employees who handled my original account have retired from work or from life; the amount of money in my account has increased by a good multiple; and the new bank manager has changed the name of my account from "Saving" to "Spending". The non-changing factor is the personal attention I receive from the management, making me the recipient of what I believe is preferential treatment, not because of the size of my purse, which is by the way very modest, but because of reciprocating loyalty to a longstanding client-customer relationship.

The HSBC, from its Hamra branch, was also the bank that extended financial facilities to Elias for the Sports City Project. *Like every prudent bank, the officer-in-charge reviewed the contract terms and conditions; assessed the political climate; studied thoroughly the project's budget, cash-flow, and cost elements; and assessed the risks of default prior to approving the financial package which consisted of cash loans and bank guarantees.* Once the facility was approved, the bank assigned one of its employees to monitor the progress of the project and report any potential performance problems that could pose a risk of financial loss to the bank. From his first visit to the project site, the bank officer and I developed an excellent business relationship based on mutual trust and acknowledgement of each other's expertise. This rapport simplified his mission during subsequent periodic site visits where I gave a briefing of the project status and he took notes; he did actually and occasionally ask interesting and intelligent questions, for which I always had the answers and explanations that served the best interests of the project. The bank officer's periodic favorable reports were catalytic in extending this rapport to include the bank's branch manager who, being the brother-in-law of one of Vicky's best friends, was inclined to allow friendship to mix with business. The flexibility in servicing the loans; the occasional increase in loan limits; and the ease in obtaining letters of credit for material orders were some of the benefits derived from this good relationship. When the Main Contractor on the Sports City project terminated our contract by default, it was the HSBC that stood by us politically and financially. The bank went as far as financing the services of the best law firm in town and the arbitration case in Paris, France.

Camille

During the last days before all Hell broke loose between us and the Main Contractor, a nasty letter found its way from HSBC to my office, signed by the new branch manager who had inherited our file and all its miseries. I still prefer to believe that Camille's appointment to Hamra-branch was in line with the bank's policy to rotate managers among its branches, and that it had no relation to our non-performing account, or to Camille's predecessor's leniency. Camille claims he gets those special vibes whenever he scans though a customer's file that spells a troubled account. He sure got those vibes that induced him to declare Elias in default, from the bank's financial standing, and to request immediate settlement of the loans. The request was rhetorical and equally impossible. Armed with a self-convinced belief in my strong friendship with the bank's higher echelon, I stormed into the manager's office at Hamra branch, ready to put up a fight for what I believed was gross misunderstanding of our solid contractual position and erroneous judgment by the novice manager. Camille was no novice, as I found out after less than half an hour through that first meeting. He was calm, serious and meticulous, had done his homework and intelligently assessed our financial situation, and possessed sound judgment with sharp instincts. Our friendship started there and then. We cut the chase of hollow bravado and went to the core of the problem: Elias, his company, and the project were in deep trouble, and the bank had shouldered a good part of the loss. Camille and I had to work together to salvage whatever could be salvaged from the financial wreck.

Arbitration in Paris

To this date Camille keeps reminding me that he never ever believed that the arbitration case between Elias' company and the main contractor would rule in our favor. He repeated that for the umpteenth time just the other day when I visited him at HSBC headquarters, where he is now serving as a training manager even though he has officially retired. Yet, during that crisis time, he helped by providing all the necessary financial information pertaining to our case which was available at the bank, and most importantly he did not oppose the bank's decision to finance the arbitration case, although he did not agree with it.

The arbitration case presented my second involvement with lawyers and rekindled my dream for becoming a lawyer myself. Elias hired a prominent law firm in Lebanon whose major partner had served few years earlier as an arbitration judge with the International Chamber of Commerce in Paris. Since my two-year contract with Elias was over, and since I had the full case history of the Sports City project properly filed in my mind's databank and easily accessible through an active memory outlet, the law firm hired me on an hourly basis to provide the facts and figures that form the backbone of the case. *Arbitration, a form of dispute resolution, is becoming more and more the contractual and legal venue for resolving conflicts between contracting parties. It is intended as a more efficient and much faster means of legally resolving conflicts than going to overburdened courts of law.* Preparing the necessary documents and legal brief then filing the case took few months. Hearing sessions, presenting arguments, and submitting counter arguments took couple of years. I got paid for the hours I spent at the law firm and was reimbursed for the cost of my two trips to Paris for the hearing; Elias and his partner unfortunately got zilch. The Arbitrator's ruling against Elias was a surprise and a major shock, and quite the opposite of the impression he gave us during the hearing sessions. Elias and many of his creditors, suppliers, subcontractors, and employees suffered financially; including me. Many went to Lebanese courts to recover their moneys, but not I. As Vicky says, after my major financial loss with Sogex, I developed a mental immunity against financial shock and formulated my own philosophical view with regards to accumulation of wealth. This psychological armor protected me when I lost most of my accumulated savings with the bankruptcy of Double Deal Pizza; later when I was not properly compensated by my friend Adel for organizing his company; and also when I lost my house in Corona to the loan company. My motto was: I earn money but will never win it; saving money is a fruitless effort; and better spend it before losing it. How many people can claim that external factors, born by pure chance, materialized at every milestone during their lives as students and resulted in cancellation of all their graduation ceremonies? I can and already have.

Unrealized Dream

The dream to become a lawyer has faded a lot but has not lost all its traces. At Kuwait airport, while waiting to catch a plane to Beirut, I met an old acquaintance from DAR Company who was taking a two-year course, by correspondence with a British university, to become a certified arbitrator. I took the email address of this university; corresponded with the concerned department and got all the necessary information and registration forms; but, my enthusiasm somehow evaporated long before the due date for registration. I am a regular reader of the Economist magazine; I scan through many of the topics in each issue; but never miss a word in the advertisement sections for offered courses. This could be a manifestation of remorse or guilt-feeling for never going for a postgraduate degree. Maybe that is the reason behind my pushing Majed to embark on his MBA program, although there is no need to push as it is on his career agenda. Mira went for her Masters in School Counseling after one year of work experience; Majed says he needs four to five years of work as his undergraduate degree is in electrical Engineering. Neither Vicky nor Mira take him seriously when he argues that the need for five years is obviously because we engineers are on a higher professional level and are thus subject to more stringent career requirements than liberal arts graduates, meaning his mom and his sister; it is one of those teasing jokes that we enjoy as a family.

While helping the lawyers in the arbitration case, I got an assignment from Adel, a classmate of mine at AUB, to manage his contracting company in Lebanon. Adel, who was well established in the United Arab Emirates as a road construction contractor, was awarded a contract to build a section of the highway that was planned to link Beirut with its Southern borders. I was free at the time; Adel was spending most of his time in Abu Dhabi; and the company in Lebanon needed proper management and control. Within three months it became apparent to me that the combination of Adel's family involvement in the company and the complexities of dealing with government officials, who represented the client on the project, were obstacles beyond my professional endurance. I parted with Adel, with traces of strain on our friendship, and accepted an assignment in Turkey as a project director on one of DAR's projects.

CHAPTER VIII: 1989 and 1999

Istanbul

Rabeeh lived in Jeddah in an apartment on the same floor as my cousin Marwan; not the same cousin who was the second bridegroom in our double-wedding ceremony. My father was the first Marwan of Monsef; other Monsefites and relatives, including my uncle and my aunt, and myself included, adopted the name and bestowed it on our children. Rabeeh and I met occasionally within our professional circles; he being with DAR, a leading consulting firm in the Middle East and the international engineering market, and I being a member of Sogex, the fast-rising talk-of-the-town contracting company which dominated the Saudi market at that time. Our friendship, initiated by professional recognition and cemented over barbeque lunches at Marwan's apartment, remained intact despite long periods of minimal contact. Rabeeh was promoted out of the local Saudi office to a director's position at the headquarters in Beirut; I left employment and focused on management consultancy, pursuing offered missions, and relocating to the country of the accepted mission.

DMM, DAR's subsidiary company in Turkey, was providing project management services to a prominent Turkish Group of companies that was developing a huge residential project, at a lakeside forty kilometers north of Istanbul. The contracting company within the Group was executing the project for its sister development company that was the client of DMM. A power struggle between the two sister companies, which had never worked together or for each other before, necessitated the creation of an operational system and procedures that defined and regulated the administrative relationships and the lines of communication among the staff of both companies. DMM's appointed director on the project had to leave half through the project, to the dismay of the Client, and had to be replaced. Rabeeh introduced me to the managing director of the Turkish Group who accepted my appointment as the replacement project director for the remaining three months duration of DMM's contract. Three months later DMM contract was extended; the Client was satisfied; and I ended up spending all of 1998 in Istanbul.

People of Istanbul

The unique charm of Istanbul is addictive. It is truly a Western community governed by Eastern culture and living in an Oriental setting which was founded on four empires. The people of Istanbul, or at least those whom I had the opportunity to interact with, on business and social levels, were very friendly, warm, and always ready to extend a helping hand. Koray started as a business acquaintance and now, ten years later, he is a happy father and a dear friend. Majed and his cousin Ramy spent a fantastic vacation in Turkey, thanks to the arrangements made by Koray; he still asks and is always happy to hear the latest news about Majed. Metin owns a ladies clothing shop in Osman Bey area. He comes from the North-Eastern part of Turkey, from a village bordering Syria, where most of the villagers speak Arabic fluently and Turkish barely, his parents included. Vicky and I walked into his shop attracted by his Arabic conversation with a customer; shop owners in Istanbul rarely spoke other than Turkish. In fact, eloquence in foreign languages was not so common; the professor of Architectural at Istanbul University, who provided technical advice on the project, needed a translator to render communication with him meaningful. Metin was happy to let Vicky translate the Algerian Arabic of the customer to the Lebanese/Syrian Arabic which he understood. In return, we became one of his very few retail customers as his business was mainly wholesale; notwithstanding the fact that Vicky's purchases for her and her female relatives and friends were often and in such quantities to qualify them as mini-wholesale.

Life in Istanbul

Vicky spent most of the year 1998 in Beirut with Marwan, Mira, and Majed who were still students; Vicky keeping the family environment intact. She made few short visits to Istanbul, and a long one with the children during their winter vacation. As for me, the requirements of my work assignment did not restrict me from making monthly trips to Lebanon, and the proximity of the apartment from Istanbul Airport was an added help. Middle East Airlines, the Lebanese Flag carrier, flew at convenient hours from Istanbul to Beirut on Friday afternoons and back from Beirut early Monday mornings. On one of these flights the name of the captain who greeted us early-bird passengers sounded remotely familiar; for me to remember a name is a major achievement and a continuous cause for embarrassment. The captain, Faysal, turned out to be a high school classmate whom I had not seen for over twenty years. From then on I tried to preplan pretended coincidence flights when Faysal was piloting, to enjoy the cockpit seat for the two-hour trip.

During their two-week compact vacation in Istanbul, the children revived in me the curiosity for tourism; an activity that had lost its glamour long time before. Their enthusiasm was contagious, and Istanbul provided a large variety of tourist attractions: historic buildings, palaces, and mosques, each with a fascinating story of its own; a nature blessed with many sightseeing places, islands, and beaches; and the round-the-clock lively restaurants that served excellent food to a clientele usually consisting of families spanning three generations. Rivalry among in-laws vanished at the richly laid table, especially when the food served was kebab in all its varieties of which Iskander Kebab was the most popular dish. We took the boat across the Bosphorus Canal; climbed the steep pedestrian road to the Orthodox monastery at the peak of the Large Island; and shopped at the Grand Bazaar till our feet swelled beyond the constraints of our shoes. That was the last vacation that we enjoyed as a family of five; it was also the last joyful Christmas for me.

UME

At a recent dinner party held by the United Nations Development Program (UNDP) in Kuwait I met the Ambassador of Turkey, who was seated next to me, and his wife, who was sitting to the left of her husband and next to Vicky. In no time I started reminiscing about the time we spent in Turkey during my first mission with DAR, in the year 1998, and the subsequent two-year mission, the second one with DAR, in 2001 and 2002. I must have overdone it with my enthusiastic detailed recount of events and places that, despite his polite invitation to visit him at the embassy, he failed to recognize me when we met two weeks later at another UN function. I don't blame him and I think I know why I subconsciously exaggerate the enjoyment we had in Istanbul; my mind is always trying to bridge the memories of these two periods and short-circuit the bad times in between.

In November of 2000 Rabeeh calls me and offers me a second mission with DMM in Turkey; his call came at a time when I badly needed a change of environment and an inducement for parallel thinking, to escape from my state of melancholy. This time the project was in Gebze, on the Asian side of the Bosporus Canal, for the Metrology Institute of Turkey, better abbreviated in Turkish as UME. DMM had the project management contract for a high-tech Lab complex, which was to be built on the campus of Tubitak, the Scientific Institute of Turkey. The World Bank, the financer of the project, required DMM to provide a Project Director of international experience, who was familiar with World Bank procedures and contractual terms; conditions that I satisfied, plus I had built a very good personal and business relationship with DMM management and staff from the first mission.

Dileg

Dileg owned several apartments including the one we rented in Atashire. Her husband is a well-established dentist with a flourishing business and down to earth personality with a docile character. Dileg decided to take over the financial management of their earnings; she bought well located apartments, furnished them, and leased them to well known companies for use by their international staff. In no time our families became good friends and exchanged visits, dinner invitations, and occasional souvenir gifts. When Mira passed through Istanbul on her way from USA to Lebanon, it was Dileg who met her at the airport and took care of her sightseeing and shopping sprees while I was at work. She is a true friend who, until now, calls Vicky on regular basis to exchange family news, inquire about our safety whenever the troubles flare in Lebanon, and to extend her open invitation to visit Istanbul and be their houseguests.

Gebze and the Local Car

Gebze is about 80 km East of Atasehir, with both towns located next to the Trans-European Motorway (TEM). Daily commuting was not so bad since the TEM was well maintained and remained accessible even during the worst snow storms and blizzards. The speed limit along the TEM, which was not posted, was rumored to be 140 km per hour. My French car felt better at speeds of 150 to 160 km per hour, a vast change from the locally made car I had in 1998. An interesting feature of the local car was the simplicity of its door locks. I was driving to work, one lovely spring morning, when the car went dead. I called the garage that DMM had made agreement with to provide me with road service when the expected failure, a fact however with unpredictable occurrence date, became a reality. The instructions delivered joyfully in Turkish, which I had them repeat few times to make sure I got them right, were to lock the car, leave a note with DMM name on the windshield, and take a taxi to my final destination. As to where to leave the car key: take it with you as all cars of that model share the same lock type and can be opened with the same key. This fascinating feature was later confirmed when Vicky and I got back into the car, which was parked along a very busy street in Istanbul, and found all the four doors locked; the windows closed; no sign of any forced break-in; and where the car radio/cassette unit was mounted only the holding-down screws remained.

Hereke Carpets

Hereke, a town close to Gebze, is where the famous Turkish silk carpets carrying the name of this town are hand woven using the old and traditional techniques. We have a small Hereke carpet, now mounted on the wall of the entrance hall in our house in Monsef, which we bought from a traveling carpet merchant in Yanbu many years ago. We got to know this merchant when he materialized at Sogex Village in Jeddah, back in 1978, and asked to display his stock of carpets for eyes to see and wallets to eventually open up. Vicky was then the president of the Women's Club and thus the organizer for such an event. His prices were good, since his overhead was only the cost of running and maintaining his van, and what I paid for the carpets that I bought was even better, thanks to the organizer of the bazaar. Three years later Vicky made the same arrangements at Sogex mess hall in Al Khobar; and the last display she set up was at the Family Entertainment Center in Yanbu in 1984.

Many of the carpets we own were bought from our merchant friend; the one carpet I did not buy is the one that enchanted me most. It was a five-meter by three-and-a-half-meter masterpiece of pure silk, depicting a floral arrangement in all the colors that have ever been recognized by the human eye. The finesse of its weaving and its floral design would not be appreciated or preserved except by hanging it on a wall along its length; our modest family home in Monsef had a high ceiling in comparison to normal construction, but still was half a meter short. Knowing the unlikelihood of ever owning a chateau with soaring high ceilings, the only setting majestic enough for such a Hereke, Vicky's advice of not buying it prevailed. The Aubusson tapestry in our apartment in Paris was the next best wall-hanging hand-woven piece of art that I took joy in looking at daily and for almost two years.

The first time we visited the town of Hereke I bought yet another book about carpets: the different types, their origins, the art of weaving, the tools and dyes used, and so on and so forth. I read it thoroughly with the same attention and intensity that I had read, before and after, books about wine, carpets, paintings, music, and other cultural topics. Unfortunately, and to my dismay, however hard I tried I never became an authority or expert conversationalist on any subject of artistic nature. It seems I am destined to be a silent enjoyer and appreciator of art, in all its forms, and only capable of sensing it rather than discussing its merits.

Hannibal

Hereke is famous for its carpets, but Gebze is famous for housing an important element in European history. Hannibal, the famous conqueror who crossed the Alps Mountains on the back of his elephant, is buried in Gebze, on the grounds of Tubitak. The grave is modestly marked and not well publicized; I came to know of it from my friend Wahid. A mechanical engineer with a doctorate degree in Fluid Mechanics, Wahid was one of the most senior employees of UME as well as my counterpart on the project. A researcher who calibrated the equipment that is used to calibrate metering instruments, Wahid trusted me with contractual issues the same way I trusted him with technical matters. This smooth professional relationship was instrumental in the management and control of the project, and in overcoming the difficulties created by incomplete design; the senior engineers of the consulting firm that was responsible for the project design had a terrible car accident and four of them were killed.

Fluid Mechanics Lab

Of the many laboratories contained within the complex, the Fluid Mechanics Lab had suffered the greatest design deficiency. With Wahid's theoretical knowledge and my practical experience, we were able to modify the engineering design to improve its function and performance. The next step was to issue a supply-installation contract, through the World Bank, and seek offers from international companies with related experience. Toni, my friend from Qatar who managed the Bechtel project in Tanzania, had just completed a five-year communication network contract in Beirut, and was recovering from a back surgery that affected his motion control system. Toni needed a motivation to exercise mentally and physically, otherwise the temporary inconvenience would become a permanent handicap. Wahid needed a competent contractor whom he could trust to build the high precision Lab, guided by an engineering design that lacked most of the construction details and specifications. Toni had tutored me on the intricacies of pipeline construction in Qatar; had shown me the mysteries of large floating-roof oil tank erection; had helped me out with the electrical works when I worked with my friend Adel before my first mission in Istanbul. I wanted to reciprocate, and so asked Wahid to invite him to bid for the project. Toni's offer was considered the most favorable by the World Bank; in two years he built the Fluid Mechanics Lab and rebuilt his strength and confidence in his motor system.

Kemal and Tuncer

On Friday at 10 AM, as for every Friday since the start of the UME project, the Contractor's project manager, Tuncer, walks into the conference room adjacent to my office accompanied by Kemal. I had not seen or heard from Kemal for almost two years, since the completion of my first assignment with DMM at the end of 1998. He was then the project manager for the Developer and, being the only Turk on site with reasonable fluency in the English language, was my weekly companion to the movie theater at the Developer's family club. Several times the theater doors were opened upon Kemal's request and we were the only audience. That did not bother me as after-work hours were very slow for both of us, Vicky being in Lebanon and Kemal's family in Ankara.

Go American

Tuncer announced that Kemal had joined the Contractor's firm and would take over the project management while Tuncer moved to the Head Office. After all, Tuncer was the company's Director and was acting as project manager to meet the Contract requirement for an English speaking manager; something that was not very common in Turkey at that time. I knew that Kemal was competent enough to satisfy the requirements of UME and DMM, and being an old friend was an added value; I approved his assignment. The first order of business was to recite the story of when I was stopped by the traffic police at a roadblock along the TEM, and the subsequent similar experience of Kemal. It was along the TEM while on my way back to my apartment in Atakoy, after watching a movie exclusively run for Kemal and I, the traffic police had erected a roadblock next to the Toll Plaza and stopped me along with every other motorist. I was advised in Turkish that one of the two headlights was off, and the policeman asked for my driving license and the car registration. My Turkish language skills were meager, however the vocabulary included many Arabic words or distorted Arabic, which when logically arranged, indicated the essence of his request. To avoid the unknown, I presented him with my American Passport, volunteered one statement, and repeated it to the limit of his endurance: No Turkish, I am American. Few minutes later, his frustration reaching sky-high, he moved aside and said the words "Go American" along with few mumbles and rumbles that must have been crude, rude, and impolite. I left and had the headlight fixed the following morning. Couple of weeks later I was stopped at the same check point, where the police had erected a road block after, what I learnt later, an incident of hit and run few kilometers before the roadblock. The policeman approaching my car suddenly stopped and started an argument with another laughing policeman who was checking the documents of another vehicle; he looked then at me and said: Go American. Kemal, who had an American accent developed through his interaction with Bechtel employees, used the same tactics when he was pulled over for speeding, on a Friday night, while on his way to Ankara; the speedometer of his Beemer was reading close to 200 km per hour. He must have been so convincing that the policeman let him go without even looking at his passport; Kemal does not have an American passport.

Learning Turkish

The Director of UME, the boss of Wahid, was attending the meeting. He liked the story and, to avoid a recurrence the outcome of which might not be always favorable, he offered to assign his Human Resources Manager to teach me the Turkish language. For the rest of my mission in Gebze, I had three two-hour sessions per week of intensive Turkish in my conference room. My understanding soared to almost seventy percent of any business or social conversation; speaking skills were enough to give the listener an idea of the message I needed to convey; and my reading abilities remained at an early elementary level. Five years later, with almost no practice whatsoever, when I tried to communicate with the Turkish Ambassador at the recent UN party, I realized that what little Turkish language skills I had existed no more.

Marwan's Recurrence

It was the beginning of January 1999, few days after finishing my first mission in Istanbul, two years before starting the second mission there, and eight years since Marwan had his operation. During those eight years we educated ourselves as extensively as possible on Medulloblastoma: A cancerous tumor originating in the cerebellum which is located in the back of the brain above the neck and is responsible for controlling body movement. Despite the confidence that the Doctors had in the complete effectiveness of the protocol treatment that they administered during the year following the operation, and the clean bill of health they willingly issued, we opted for the yearly MRI to ease our cyclic anxiety build-up and to bring it down to an acceptable restart level. God rested on the seventh day; we lost our short-lived rest on the seventh year; and my memory of that horrendous day is coming back to me while I am writing on this seventh day of the seventh month of the year seven after the first two millenniums. The MRI revealed traces of re-growth of the tumor, to the amazement of the Doctors and to our disbelief; possibility of recurrence was supposed to be nil and we had augmented the odds against it by daily prayers. All the specialists we consulted recommended stem cell transplant as the most promising option for possible recovery, which we agreed to despite it being still a clinical trial or in the experimental stage.

After quick over-the-phone consultation between Dr. Joseph, the Surgeon who operated on Marwan and was at the time working in Lebanon, and Dr. Bedros, the Oncologist who gave Marwan his post-operation treatment, arrangements were made for Marwan to be admitted at Loma Linda for further tests and most likely stem-cell treatment. While making our travel arrangements we found out that Marwan's Lebanese passport had expired. We had been using our Lebanese passports to enter and exit Lebanon only, as a measure of avoiding the long wait to get a visa at the airport; outside Lebanon we were Americans with American passports. The Ministry of Interior high official, who was asked to assist in getting Marwan's passport renewed in one day instead of the normal seven days, advised me that the Lebanese Identity Card is a sufficient document to enter or leave

Lebanon, along with the US passport. Since then we took it on ourselves to pass this information to all our friends who had opted for nationalities other than the Lebanese, in order to ensure a safe and stable future for their children.

For almost seven months Marwan was in and out of Loma Linda hospital, with me at his side pretending bravado that I did not possess. Khaled was there for us, offering his house during the out-of-hospital days, and moral support that we all badly needed. The lesson I learnt from this ordeal is that there is no limit to the human being's endurance. Praying lead to hoping; fear and worry boosted the will to resist; and the meaning of faith suddenly became simple and comprehensible. It was then that I recalled a conversation I had, at the age of ten when curiosity prevailed, with a wise religious Bishop of our church. I asked him how he could prove or at least convince me of the existence of Paradise. He said that faith does not go through the channels of logic and scientific facts, but is the product of pure acceptance of ideas and concepts that give meaning to questions that have no answers. It is a combined mental and emotional state where the unknown is described to make it graspable and thus eliminate the fear associated with it. A conversation went as follows:

Bishop: "Do you, Walid, behave in the manner that your parents have instilled in you?"
Walid: "Yes."
Bishop: "Are you aware that this behavior is in line with the teachings advocated by the Church and other religions, the same teachings that promise you of Paradise?"
Walid: "Yes I am aware."

Bishop: "Will the existence of the promised Paradise or the lack of it affect or change your behavior and convert you to a liar, thief, or cheater?"

Walid: "No, not at all."

Bishop: "So, wouldn't it be better to believe in Paradise as a reward for good deeds and honorable behavior, and if it exists you benefit in the afterlife, and if it doesn't it will not change your life while you live?"

Walid: "I guess there is no harm but only potential benefit in believing."

Bishop: "That is Faith; Walid you can now say you are a believer in Paradise and any other religious teachings you come across."

In June of 1999 Vicky joined me in California. In July Marwan was admitted to the children hospital in downtown Los Angeles for Stem Cell transplant. In August his lungs went into shock from post-transplant complications. On 12 September 1999, few days after Lama and Amin, my sister and brother, joined us for moral support, and after almost a month of being in coma, Marwan peacefully passed away. Two images keep popping up in my mind, especially when I am at an emotional low point: Marwan, few weeks before the transplant, announcing for the first time in his life that he was afraid; and the second image when he was in coma, his eyes so peacefully following me around, bidding me farewell and encouraging me to accept the inevitable. He gave in before I lost hope and before I consented to the Doctors' recommendation to let go and get him of the life-support machines.

CHAPER IX: 1999 to 2012

November 4, 2007: Today is Marwan's birthday, a day where our emotions soar to the highest level of fragility and tears are held back by adhesion for as long as it takes gravity to activate its down-pouring force. Vicky and I, even here in our apartment in Kuwait, remember this day with the usual lit candle placed in front of Marwan's framed photo; a silent prayer; mental communication with the past; and blowing off the modest flame of a single small candle crowning a birthday cake. I don't know if time will blow off the fire that burns in my heart, a fire that will not fade despite the slow fading of the memories that bonded Marwan to the family. Mira is here in Kuwait and Majed is in Dubai. We have not heard from them today even though they call almost daily. I am sure they remember this day, each in her and his own way; their silence is to spare us the heart-bleeding triggered by remembrance.

The two years in-between the two missions in Istanbul, one year in California and the other one in Lebanon, have become distant and vague memories resisting any attempts for clarity or awareness. They are defined by three events on a linear timescale: Leaving Istanbul on 22 December 1998, Marwan's death on 12 September 1999, and returning to Istanbul on 24 November 2000.

By the time UME project was over, two years had passed, the hurt had subsided to bearable levels, my limited skills in the Turkish language had developed, and I was ready for another mission. What I was not ready for or counting on was an assignment in Saudi Arabia. DAR had been awarded a management contract for the development of Al Khafji oil city, near the borders of Kuwait, by a company owned jointly by Saudi Arabia and Kuwait. The first phase of DAR's contract was a three-month period during which the concept design was to be completed by the contracted Engineering Firms, the alternative development proposals were to be defined, the Construction packaging was to be determined, and the project control and document control procedures were to be established. My scope of work as per my contract agreement with DAR was to manage this first phase as DAR's Project Director. Renewal of my contract was subject to few conditions of which only DAR's contract extension to cover the design and supervision phases was met.

DAR is well known for its efficiency and speed in mobilizing people and services to meet the needs of any project in any country or any city. Establishing a DAR office, a project office, and accommodation facilities for the project staff was particularly easy in Al Khobar since as a company it had existed on a large scale and for many years in Saudi Arabia. But Al Khobar that met me on the 3rd of January 2003 was much different from the one we lived in two decades earlier. More highways crisscrossed the trio-cities, Dahran-Dammam-Al Khobar; many high-rise buildings mushroomed all over the place; a large Western style shopping mall was the recent sanctuary from the heat and humidity; but only very few Westerners, by nationality or by culture, resided in Al Khobar. Years before, when the Royal Commission was building the industrial city of Jubail, the housing compounds in the trio-cities were exploding at the seems with Western, Eastern, and Middle-Eastern nationals. The area was like a beehive with people of all ages, colors, and shapes buzzing at the Souks, the beaches, and the restaurants. With Jubail city completed, residential as well as industrial, it seemed that Al Khobar had faded out and gave in to the new modern development only thirty minutes away.

My work was very challenging, overwhelming, and consumed all my waking hours. Vicky was supposed to join me only if and after renewal of my contract beyond the first three-month period; no distraction could be afforded to meet family needs during this initial period. Neither Vicky joined me nor I stayed beyond the original duration; the Client's representative and I did not see eye-to-eye on how the project should be managed. He had the political power which he insisted on utilizing to enforce his point of view, and I had the professional expertise and business ethics to reject unwarranted or erroneous interference. There was no way I could or would accept his considering and treating my Document Control Manager, an engineer with a PHD in management, as a secretary whose job was to manage the filing system. The Client did not need to compromise and I had no reason to especially that the UNDP had offered me an assignment in Kuwait and I had already accepted.

Kuwait

Marwan, my longtime friend and member of our Group Seven of IC, called to say that he had passed my email address to the UNDP program coordinator in Kuwait who was looking for development experts. A few days later Samir, a person who carries the same name as several of my friends and relatives but whom I had never met before, sent me an email with terms of reference (TOR) for a consultant needed by the Ministry of Planning in Kuwait, under the Cooperation Program between the State of Kuwait and the UNDP. I sent him back my curriculum vitae and expression of interest in the mission. A month later, I had negotiated a financial package, signed a preliminary agreement, and flew to Kuwait; it was the fifteenth day of June of the year 2003.

UNDP's Mission

Within two days I had rented a furnished and serviced apartment, leased a car, opened a bank account, and was ready to start work on my six-month consultancy contract agreement; Samir's well established contacts made all that possible in record time despite my being on a visit visa and a short term contract. My office was on the third floor of the Ministry of Planning; my primary assignment was to support the Consultants Department within the Project Development Sector; my secondary duty was to provide technical advice to the Undersecretary of the Ministry; and my mission was the transfer of knowledge to local cadre.

The "Committee for the Selection of Consulting Firms", better known as The Committee, with members representing certain ministries and authorities, was established in Kuwait with a mandate to select the most suitable consulting firms, for most of the Government development projects, and to authorize the concerned Government beneficiary to conclude a consultancy agreement. Registering consulting firms at the Ministry of Planning, inviting consulting firms to submit offers for the various projects, review and evaluation of these offers and providing a final recommendation of the most suitable consultant are all the basic functions of the Consultants Department. The Government beneficiary initiates a request for hiring a consultant for a specific project, which in most of the cases is a development project, and provides verification of budget allocation to cover the project management, design, supervision, construction, and construction management of such a project. Occasionally, a request for a feasibility study or administrative reform plan or strategic plan is requested. The beneficiary also prepares the terms of reference (TOR) for its project to identify the scope of work of the project, the scope of work of the consultant, any special conditions it intends to impose on the winning consultant, and the details of how, what, and when to submit the technical and financial proposals.

Consultants Department

The engineers of the Consultants Department prepare a list of suitable consultants to be invited, who are selected from the data bank of registered consultants at the Ministry of Planning, and after agreeing on a short list with the beneficiary, issue invitations for submission of proposals. The received technical proposals are evaluated by a joint sub-committee of pre-appointed members of the beneficiary and the Consultants Department, and the results of the evaluation are raised to The Committee for selecting the chosen one. The process, at face value, seems straightforward and simple; it is not. *The TOR, which is the basis for the proposals, should contain comprehensive information as to the requirements of the beneficiary. It needs to be in a unified format to deliver a clear understanding of these requirements to the consulting firm, and should be linked to the evaluation process to ensure a fair and equitable evaluation of the proposals by the members in the sub-committee. The evaluation criteria should be standard and must follow internationally recognized and accepted systems to ensure the hiring of the most appropriate consulting firm for any particular project. The bidders should be selected through a process that matches experience and expertise with the scope of work; ensures equitable distribution of projects among local firms based on their work load; allows the choice of international consultants that would provide an added value to the project and the transfer of knowledge to the staff of their local partner; and takes into consideration the past performance on Government projects with respect to quality of work, timely production, and completeness of product. Evaluation of proposals should be carried out by qualified staff, of proven integrity, with each member working independently; individual results are then combined for final scoring. The financial proposals should also be evaluated and their scores added to those of the technical proposals, using a pre-agreed upon weighing system that reflects the beneficiary's preference of quality over cost, or cost over quality; the complexity or simplicity of the project is the main deciding factor.*

Natheer

Natheer joined the UNDP as an expert consultant for the Consultants Department of the Ministry of Planning less than a month after I had started. We shared the same office; the same quest for knowledge; equal enthusiasm to simplify and yet improve on existing procedures and the modus operandi; similar views on many political and cultural issues; and matching belief in our abilities to transfer knowledge and to contribute positively to the performance of the staff, the department, and the ministry we were part of. An architect, university professor, researcher, and ardent writer Natheer indirectly prompted me to start writing this book; his open enthusiasm is well acknowledged. Together we either developed or refined existing registration procedures of international consulting firms; classification and ranking system for international and local consultants based on their manpower and financial capabilities and past experience; proposal evaluation system based on World Bank procedures; performance evaluation plan for consulting firms to be adopted by beneficiaries; and internal operating procedures based on a control and monitoring role for the Consultants Department rather than equal participation with the beneficiaries in the evaluation process. Natheer went back to teaching; I moved from the Ministry of Planning to another mission with

UNDP in Kuwait; and now, more than four years later, much of what we developed is being applied, yet a lot remains to be done.

Friends in Kuwait

The first person I called in Kuwait was Houry. His wife, whom I had never met before but we knew of each other, volunteered to show me around and to help in searching for an apartment. Since then, we have met occasionally and always Houry and I reminisce over the past and overwhelm her with our shared memories that go as far back as memory could go. When my parents moved to Hamra Street in Beirut, Houry's family were our neighbors; they lived on the second floor and we were on the third floor. Houry, the youngest among his siblings was my classmate from elementary school through graduation from AUB Engineering School; his sister, the one next in line age wise, was my sister's classmate for the same duration. His father took turns with my father to drive us to Ahliah School whenever our driver, Joseph, was not available. Joseph was more fun as a driver; he did not differentiate between the road and the sidewalk when we were running late and were stuck in a traffic jam, and red lights were not taken seriously by him. Almost every summer vacation Houry and his sister Huda would spend few weeks in Monsef at our place, behaved like family members, and developed a love for Monsef and Monsefites that keeps their dream of retiring in Monsef alive yet far fetched.

Michelle

I am not the only driver who needs Vicky as a copilot and navigator, without whom I would end up in Tokyo, as my children say. Michelle is even worse when it comes to lack of sense of direction and equally relies on Vicky. She refused her son's offer to buy her a car road-map finder, based on Global Positioning System, claiming that Vicky is more reliable, responds instantly to changes in traffic conditions or missed highway exits, has a memory bank of updated detailed map of Kuwait with all the shops clearly indicated, and communicates information verbally and with frantic sign language and waving hands. Samira, a professional buyer of clothes for her own use or for the pleasure of shopping, fully supports the driver/navigator agreement of these two friends as an insurance against wasted shopping time.

Samira

Samira is a good cook of Lebanese food and especially "Malfoof", the stuffed cabbage that takes time, effort, and skill, which she claims is prepared only for me as token of being a favored friend; Ghassan, her devoted husband and perfect host, plays along acting as the jealous husband who is deprived of this special attention. She was the second person I called when I first came to Kuwait since, per Lama's advice, she is the Lebanese "Mokhtar" in Kuwait, and the encyclopedia of who's who among the Lebanese community; an excellent source of information for anyone wanting to quickly integrate socially. A Mokhtar is the lowest civil servant, in terms of the Lebanese government employment hierarchy, who is elected by public vote to represent the smallest administrative unit; a village or a confined geographic area within a city. His primary responsibility is to personally know every citizen or resident within his area of jurisdiction, and to officially attest to that whenever it is required as part of government formalities. He is the only authorized person to handle the paperwork for birth registration, for the officially deceased, and for obtaining civil identification cards. Unfortunately, the system of communication among government agencies in Lebanon and many other countries that I have lived in is rated poor to non-existent; it took me five years to find my renewed Lebanese civil ID card, which had been placed in an envelope in a metal cabinet, in the government civil service department offices, and had been dormant there for two years. The ID card of the Mokhtar's wife, who happens to be a cousin of mine, was among the twenty or so dormant in that envelop; he had no way of knowing as the official advice was: "not yet issued". A friend of mine and former employee from the days of the Sports City Project used his connections at the Ministry of Interior to find the whereabouts of my ID card. Vicky, Mira, and Majed waited for almost ten years for news about their ID cards; gave up; reapplied for the umpteenth time; and, eventually Majed got his ID card when it became mandatory for voting in Parliamentary elections. Mira is still waiting for hers, still hopeful after applying for four times.

Golden Girls

The Golden Girls, Vicky's close group of friends from elementary school through university studies, were represented in Kuwait by Hossah. Going through the photo albums that Vicky treasures from her pre-teen days you find Hossah and Vicky climbing trees, riding bicycles, in colorful clothing at the folk dancing festival at AUB, and munching on junk food with other members of their group. The concept of the Golden Girls is not much different from my Group Seven except for their more frequent get-together parties, sometimes with the husbands but more often girls alone. We, husbands, prefer the parties held at home rather than in public places like restaurants. The noise, loud laughs and simultaneous conversations that they generate, surrounds them like a steel sphere shielding them from their surroundings; they are genuinely oblivious to complaints or discomfort that might arise from other customers.

Hossah

The other day we were at Hossah's place having a light dinner of fish kebab and salad; Rakan, Alec, and Lucy were the only other guests; not that we consider ourselves as guests, in the sense that we are required to observe social protocols, but more like family members. Such small gatherings, often taking place when Hossah gets back from a business trip, as well as more formal and informal dinner parties, always at her place, are where we met most of our Kuwaiti friends and many interesting acquaintances. The evenings there are always enjoyable and especially so when we play "Cote by six"; a Kuwait card game, played by two teams of three partners each, that was easy for me to learn and soon became the most sought after partner. Little did the other players know that we Monsefites are born with a deck of cards in our hands, and learn how to shuffle the cards before we even start walking or talking; exaggeration notwithstanding.

Equally enjoyable are the intelligent conversations that go on, easily and smoothly moving from one subject to another; ideas are exchanged and personal opinions are expressed; local, regional, and international news are analyzed; economy, politics, social events, and humanitarian acts are addressed with the same ease; race and religion are rarely mentioned as the group is non ethnic and comfortable with the privacy of the beliefs and religious practices of each individual. Fanaticism, blind loyalty, uncompromising stands, and unequivocal opinions somehow never come up in our discussions; no anger and no argument for the sake of argument. How so different from the social gatherings these days in Lebanon where an opinion is stated as an eternal law and disagreement is considered a heresy.

Kuwaitis & Lebanese

Regional and international news, regularly discussed in our gatherings, are dominated by the political situation in Lebanon; will there be an election of a new president or not, and will it be carried out in a timely manner to avoid "vacancy" in the government apparatus. How such vacancy is prevented or otherwise realized, in accordance with the Lebanese Constitution, is a debatable issue that has given birth to hundreds of political opinions. A first time guest on Lebanese TV stations is a political commentator; twice appearing on the same TV station becomes an analyst; frequent interviews on more than one station is a promotion to political experts. Logical analysis and accuracy of information are secondary to apparel and background setting. Kuwaitis, being not less interested in the welfare of Lebanon than the most devoted and loyal Lebanese, follow up closely these political debates and have succeeded in filtering out the sensible information from the ridiculous fanfare. The closeness that Kuwaitis feel towards the Lebanese is probably related to a similar inherent nostalgia for sea-going trading, shared love for worldwide travel, equal hunger for modern trends in appearance and behavior, their adoption of a democratic government system that is often misused by the ruling members of their societies, and their insatiable need to own property outside the boundaries of their countries. It is said that the number of residential units owned by Kuwaitis in Lebanon are at the rate of one unit per every two Kuwaitis; a proof of this could be the annoying fact that direct flights between Kuwait and Beirut, which are provided by several airlines, are normally full and always overbooked on special holidays.

Tragic deaths by unnatural causes that have occurred recently in Lebanon, when several prominent Lebanese leaders were victims of vicious acts attributed to political affiliations, are promptly analyzed at Hossah's social gatherings. Theories and counter theories are thrown around, reasons and causes are evaluated in depth, winners and losers are assumed and presumed, local and international alliances are introduced in the analysis, and at the end the only conclusion we all agree upon is that it should stop immediately by any and whatever means. Vicky and I would leave with offered condolences from our Kuwaiti friends and acquaintances, who shared with us the grief and anger of all the Lebanese community in Kuwait.

Economic Situation

The recent soaring price of crude oil has dominated all other issues in our social debates on world economy and its cascading effects on local life in Kuwait. Underground crude oil is considered by Kuwaitis as a generous gift from the Creator and belongs to all Kuwaitis and thus they have the right to its revenue, to be shared among the individuals and the government spending needs. They have a right to reap the profit from a barrel that sells at almost one hundred US Dollars when it costs a fraction of that to produce. The regular citizen does not want to be bothered with distracting and complicated issues like today's purchasing power of a $100 bill compared to ten years ago; the factual rapid depletion of this limited natural resource; the importance of reinvesting this wealth to generate sustainable future income; the significance of diversified sources of national income and non-reliance on only one product; these are all minor inconveniences to be ignored. Fortunately, there are many Kuwaitis, among them our social friends and many business acquaintances, who care about these important issues; the future would tell if our exchanged ideas were theoretical or, by being successfully implemented, are practical.

Political Situation

The future is now fogged with regional political conflicts that have left the World boiling with diplomatic and military activities, with Western, Eastern, Far-Eastern, Middle- Eastern, and maybe Far-Western interests thrown into this stew. The Palestinian issue has added a new dimension other than Arab-Israeli conflict; the dispute between the internationally recognized Palestinian Authority and other Palestinian factions that are accused of usurping power in parts of the presumably free-rule area. The Iraqi-Iranian war has given way to Iraq versus the rest of the World and Iran versus nearly the same World players. Iraq has invaded Kuwait; Kuwait was liberated by the US military machine; Saddam Hussein remained in power; economic sanctions were imposed; Iraq's alleged ownership of weapons of mass destruction prompted international USA-lead forces to invade Iraq; now Iraq is a mess; and after few years of internal war and thousands of war casualties the weapons of mass destruction are yet to be found. Iran is in the limelight for its nuclear program; the West is worried that it could be directed for the production of nuclear weapons, Iran claims it is for peaceful causes, and finding the truth is mighty costly. The origins of current World terrorism are being attributed to the Arab region and to Arab nationals; where it comes from and why it has gained momentum have been subdued by the 9-11 attack on the World Trade Center in the USA. The Middle East region is accused of harboring religious fanaticism and religious extremist, and of exporting religious intolerance to the rest of the World; the Arab World remains guilty until proven innocent, a concept that contradicts liberty and freedom for all. This we all agree on when we discuss politics.

American Charity

Many members within our social group are American citizens, including Vicky and I; some by birth, others by immigration and naturalization, and few through marriage. What we all share is the rejection of branding Americans as colonialists, imperialists, and heartless people. The concept of generalizing and non-distinguishing between politics and human behavior is a common malady. I always argue that the ordinary American people are among the most hardworking, studious, open-minded, and generous people in the World. Look at Silicon Valley; check the flag on the moon; look through the Hubble Telescope; follow-up technical development; count the number of centers of higher learning and their associated research institutions; and read the latest global sales figures of Coca-Cola, Pepsi-Cola, Pizza Hut, and McDonalds. When it comes to contributing to charities, our personal experience with the Ronald McDonald Foundation is a best example for unsolicited, unconditional, and humane generosity. When Marwan was admitted to the Children Hospital in LA, the social support office of the hospital directed us to the McDonald "Home Away From Home". It is a nearby house that functions like a home large enough to accommodate several families at the same time, with each having its own sleeping quarters, and all sharing common dining and cooking facilities as well as the reception room, family room and entertainment room. The home is managed by volunteers; food, cooked meals, cleaning material, kitchen consumables, newspapers, magazines, and even long-distance telephone calling cards are contributions from nearby companies or corporations. The Foundation provides the house and all its maintenance and upkeep costs; the full furnishings are donated by private families and individuals; and residing families of ill children provide a loving and caring environment to support and help each other, with no monetary obligation regardless of their financial situation.

Contract Renewals

My first six-month contract through the UNDP was renewed in January 2004 for an additional one year. Then it was renewed successively three times: twice with the Ministry of Planning as the beneficiary and the last one, ending in December 2007, with the Higher Council for Planning and Development, a new beneficiary headed by the Prime Minister; all this time in the State of Kuwait. During the two-month period prior to the last renewal, while waiting for new formalities to be completed, I was approached by Ramzi to activate his management company in Lebanon and be its General Manager; an offer was negotiated and an agreement was reached. Our friendship, mostly business based, started while I was with the Ministry of Planning when we found out that he had lived in Corona, California, one block from my house; at the time we were complete strangers. A year earlier I had helped him in establishing his company in Beirut, and introduced him to some of my influential friends there who could be beneficial for launching his business. Vicky packed our personal belongings and preceded me to Lebanon; it was the summer vacation anyway when all expatriates in Kuwait, especially the Lebanese, travel to Lebanon to enjoy the more moderate temperatures. I flew to Beirut on the last day of November 2006; was greeted two days later with the erection of the tent city in downtown Beirut, another manifestation of political protests; reserved a seat on a flight bound for Kuwait the day after; and signed my new contract with the Higher Council for Planning and Development within the same week. Vicky waited for Mira and Majed to leave Lebanon at the end of their vacation, and happily joined me in Kuwait.

The interesting part of my work with the Ministry of Planning was more my participation in special studies, as a representative of the Ministry's Undersecretary, than my day-to-day technical support to the engineering staff of the Consultants Department. I was appointed as a member of the committee that was established by a Prime-Ministerial decree to assess Kuwait's need for a new major seaport and the best location for it; the final decision was more political than technical. Other mega projects and studies included the bridge that was planned across the Bay of Kuwait, which could not be built on BOT (build, operate, and transfer) basis as cost recovery was not feasible; the plans for new cities to accompany future development and to accommodate a growing and flourishing population; the feasibility for a railway and metro systems for Kuwait; and the development of e-government operations for most of the ministries and government authorities.

Professionally, Kuwait gave me the opportunity to be involved with review of Strategic and Development Plans, Master Plans, and Administrative Reform Plans on a national level; it is the reward that I have sought to culminate thirty five years of work in the engineering and management professions.

On a personal level, Kuwait is the venue where I am to receive a most cherished gift to elevate me to a new level of joy. Knowing Vicky and her marrying me almost 33 years ago raised me to a high level of happiness that was maintained by the merger between our souls; she gave me a life that I would gladly live over and over again for as long as I am alive. A higher and different plateau, which I honestly believed would be the highest, was attained with the birth of our three children, who equally share my heart and my mind. What I was not prepared for is the upward leap on the ladder of ecstasy and elation when the nurse at the Royal Hayat Hospital pointed at the newly born baby behind the windowless glass wall; a chubby gift from God, delivered by Mira on the 1st day of May of the year 2008.

This is Kiira, my first grandchild who saw the light on the birthday of Vicky, but I selfishly consider her as a gift to me, notwithstanding my birthday being in December.

Kaila's Birth on 7th February 2014

I must have dozed off or went into a mental trance; for how long I could not first tell. Looking at the progression of the scenery flowing by the window through which my journey is reflected, it seems that I did not miss much; flashbacks of snapshots, strings of imprints and pictures in motion with increasing vividness clear my head.

The train is still chugging along, slower here and faster there, uphill, downhill or around a bend. If it were not for the jostling and rambling to remind me that the train is in motion, I could have bet on somebody else's life (I don't take chances with mine) that life on the outside was moving backward rather than my days inching forward.

It is common knowledge and accepted fact that every life journey will come to an end. Unless a God-granted miracle occurs in medicine, science, DNA research or whatever, in the very near foreseeable future, my trip will be over and I shall reach my final station; the possibility of falling in a mental condition that renders me oblivious of my life journey does not escape me.

Kaila's gentle screaming brings me back to reality. She wants to be fed again; two hours of sleep and a non-proportional poop, the other two functions that a 14-days old baby is limited to, were successfully executed. Kaila, the awaited first child of Majed and Maria, was born on 7th February of the year 2014 weighing almost 2.9 kg and 47cm tall; noticeably less than her measurements two weeks later. She is younger than her cousin-sister Kiira by 5 years, 9 months and 7 days. Kiira, who is absolutely engrossed with her ability to transform her eloquence into written words, decorated with multi-color drawings, sent Kaila a special letter, quoted verbatim:

Dear Kaila I am so happy that I am your cousin sister your lips are Chubby and look so Beautiful and adorable. Maria Majed are yar Mom and DaDDy. Love Kiira

HEISCO

Kiira was two months old when, due to the delay in renewing the cooperation agreement between the UNDP and the Kuwaiti Government, Vicky and I decided that it was time to move. Lebanon, specifically Monsef, seemed to be a tempting locale for an early retirement. Eight weeks later, Mira and Omar's summer vacation, spent in Monsef, was over and they had to go back to Kuwait. Kiira was going back too, but not without me. I accepted an offer of Contracts Manager with HEISCO, a Kuwaiti publicly-traded company; and, Vicky and I moved back to Kuwait.

Through the purchase of a large block of HEISCO's shares, a group of companies had gained control of the board of directors and initiated a major overhaul of HEISCO and its subsidiary GD. HEISCO was uniquely placed in the business sector of Kuwait. It had the only shipyard in the area, occupying a large part of Kuwait's primary seaport at Shuwaikh; the only offshore and onshore marine operation for dredging, reclamation, shore protection, port construction and other related works; a highly specialized and internationally certified fabrication facility for pressure vessels and other equipment for the oil and chemical industries; and an oil and gas construction and maintenance operations, with offices located in

Mina Abdallah, focusing on pipeline and tank construction and industrial maintenance.

The newly appointed CEO, a friend of mine who had been assigned the tasks of restructuring HEISCO and its transformation into a profitable company, asked for my support. While enjoying a Cuban Cigar at his beach house in Tabarja, not far from Monsef, he posed a single question: are you sure you want to be in Lebanon while Kiira is in Kuwait? A week later I ended my early retirement that had survived for merely 60 days, mind you devoted fully to enjoying my time with Kiira, and accepted the offer of Contracts Manager. My mission was to help in the turnaround of HEISCO with particular emphasis on the contractual and legal sides of the business.

HEISCO's *organizational philosophy is based on interdependency among functional units that are segregated into two groups: Operations and Services. Project and construction management and execution of works on site are the responsibilities of Operations. Everything else falls under the headings of Support and Services: Business Development to secure clients and projects; Proposals to price and win tenders; Safety and Quality Assurance to ensure healthy and safe work environment; Project Controls to monitor time and cost and advise on corrective measures when the project derails; Contracts to handle all contractual and legal aspects, before and after contract award; Finance to manage revenues, costs and hopefully profits; Human Resources and Administration to take care of the manpower element; Equipment Division to secure the machinery and equipment as required by Operations; and Procurement provides the materials and external services. In short, projects are secured and handed over to*

Operations with all the necessary tools, guidance, monitoring, control and advice with one obligation: manage the execution of the works to complete on time, to the required quality and safety standards and make profit. There is another organizational system that treats each project as an independent profit center where all the Support Services are contained within the project and are, in addition to execution, under the control of the project manager. Each of these two systems has its own supporters, defenders and advocates.

One aspect of restructuring HEISCO was the establishment of a Contracts Department as a separate functional unit, a mission with five objectives:

1. Defining the role of the Contracts Department, its goals, and its mission.
2. Creating an organization that could efficiently and professionally meet the needs of the various business units of HEISCO and GD; and accordingly recruit Contract Administrators with appropriate qualifications.
3. Develop the inter-department rules and regulations; inter-relationships among the contract administrators; training and learning requirements to enhance the department's professional output; and working tools to improve efficiency and productivity without compromising on quality.
4. Develop procedures that define the relationship between the Contracts Department and the other business units.
5. Implementation of these procedures, regulations and rules.

The Contracts Department is a service provider; as such, and being a newly introduced concept, it was important to promote it within the company with delicacy and diplomacy rendering it a service that is sought after rather than an imposed one. It is the usual balance between leniency and firmness, between flexibility and rigidity and between success and failure. The Contracts Department, in few months after inception, was recognized as a one team of serious and studious professional service providers; the Department passed every yearly Quality Audit with flying colors earning a "smiling face" and not a single Non-Conformance Report (NCR), or a "sad face".

My contribution to the restructuring of HEISCO extended beyond the establishment of the Contracts Department and managing it. I volunteered for the role of advisor and supporter to the mangers of the Business Units' managers and the upper management of both HEISCO and GD; a role that gradually consumed more than fifty percent of my time. *For an adviser, the easy part is giving advice. The difficult part is getting others to seek your advice; to respect it; to accept it; to benefit from it; not to be threatened by it; to recognize that it is delivered free from any ulterior motives; and to come back for more. It goes without saying that the advice must be founded on solid, multi-faceted and diversified experience, and it should be delivered with an aura of maturity and a non-contending manner.*

Contracts Management covers but is not limited to contract formation and contract administration. The former deals with all contractual activities prior to concluding a contract agreement, which is a legally binding document, such as: Review of contractual and legal terms of tender documents; preparation of contract agreements (Rental Agreements, Service Agreements, Consultancy Agreements, Agency Agreements, Cooperation Agreements, Joint Venture and Consortium Agreements; Pre-Bid Agreements etc.); developing procedures for contract administration; and developing and maintaining monitoring tools: forms, formats and logs. The latter provides direct support to the project team executing *a concluded contract to ensure compliance with all legal and contractual requirements of the contract and all sub-contracts; vets correspondence of contractual nature, an important tool that could work for or against the interests of the project; and handles all claims by or against the project.*

Contracts Management is involved, in addition to contracts at the pre and post award stages, with the overall management of a company where such management directly or indirectly affects the performance of the projects. Company policies and procedures, organization structure, performance of support departments, relations with clients and other business partners, management of company resources (manpower and equipment) and company strategic planning are aspects of the overall company management.

Kiira's First Year

The Head Office of HEISCO and GD, where all support and service departments including Contracts Department (CAD) are located, is in Shuwaikh; so was my main office, where I spent five days a week. On Mondays I worked out of my office in Mina Abdallah and devoted my time to advise and support the Oil & Gas business unit: Business Development, Proposals, Construction and Maintenance Operations, Fabrication and Trading. CAD monthly meeting was held on Monday since the majority of the Contract Administrators were seconded to the Oil & Gas business unit and were located in Mina Abdallah as well.

Each Contract Administrator has two functions: prime function to support the business unit he is assigned to and the secondary function to assist other Contract Administrators whenever the need arises: to cover up for each other during leaves, absences and work overload. The secondary function required a team spirit: continuous communication and cooperation among the Contract Administrators, support and mostly the sharing of experience and expertise. The purpose of CAD monthly meeting is to reinforce the One-Team spirit and to exchange lessons-learned from interesting cases and issues; more importantly, to promote camaraderie.

Choosing Mina Abdallah for the monthly meeting was for my personal convenience. Kiira was four months old; she lived, of course with her parents, ten minutes away from my office in Mina Abdallah; Mira and Omar got home from the American School of Kuwait (ASK) at five in the afternoon; that gave me one to two hours, depending on how well I managed my work, of Kiira's full and focused attention. I monopolized her giggles; her smiles, which I am convinced, were more radiant when bestowed on me; and her baby language, which I understood perfectly well and could reciprocate meaningfully. It must be obvious that soon enough Monday was not sufficient for my business in Mina Abdallah; I needed Wednesdays and sometimes Thursdays as well. Fortunately, our offices in Mina Abdallah are efficiently interlinked with Shuwaikh Head Office; geography was irrelevant. Abou Al Abed's geography was a bit different; recounting his two-week tour of Europe by car, he said: we left Beirut at noon time and by five we had dinner in Istanbul. Then we drove to Paris to catch an early movie before retiring for the day in Madrid. "Abou Al Abed, do you know geography?" "Of course," he replied, "we stayed there for two days on our way back to Beirut."

August 2009, after spending the summer in Monsef, Mira, Omar and Kiira moved to Dubai; Mira to the position of elementary school counselor of DAA, Omar as the high school IT instructor, and Kiira as Nursery half-day student. Monday at

Mina Abdallah was again more than sufficient for my work there; then every other Monday; and finally one Monday a month when CAD meeting is convened.

Almost three years later, June 2012, Vicky and I decided to relocate to Monsef; this was a final decision. Reasons: (a) We had been living all over the world for 41 years; I was close to 64 years old and Vicky had reached her maximum admissible age of 29. (b) We wanted to start enjoying our house and everything that made it a home.

Monsef Home

A very common social debate revolves around the issue of family stability: which is better, to raise a family in a single well defined physical and social environment or to establish residence in various locales around the world. Our family opted, by choice not by force, for the latter; so far we have lived in 30 houses, 15 cities and 9 countries extending over 3 continents. Did we provide stability for the family; we believe yes. Socially, we interacted with multi-ethnic, multiracial, forward-thinking, open-minded, politically and religiously tolerant groups of people; you find these people, if you so wish, in every community around the world. For our children's education, we enrolled them in an American international educational system that maintained consistency and progression in learning irrespective of geographic location. However, the most important factor was our ability to embrace the uniting family values and norms: love, support, reliability, respect, tolerance, generosity and ethical behavior.

Vicky and I, to be honest it was mostly Vicky, labored to make a home of each of these 30 houses. Success was achieved to a certain extent; but, they all lacked the element of emotional social roots. After 17 years of nomadism opportunity presented itself.

Late 1987, Naji decided to sell his ancestral house in Monsef, a two story non-homogenous old villa, after failing to convince "Aamty" Hanneh to sell him an adjacent small lot of land; one corner of the house was built on Hanneh's land long before and now the laws prevented him from carrying out essential house renovation and expansion without acquiring it. Naji gave me first preference to buy the house since his forefathers and mine had been neighbors for generations; also he was convinced that "Aamty" Hanneh would readily sell me this lot if I ask. I bought the house and Hanneh's lot; Vicky's parents moved in the summer of 1988 after the situation in Beirut became intolerable for an elderly couple; and, Vicky, the children and I moved to California, USA, in August 1989 to establish residence and claim the American citizenship in due time.

Not until mid-1996, after relocating to the Middle East as Lebanese-Americans, did we embark on the renovation of the house. Partial demolition, repairs and re-building had to be meticulously programmed to allow the use of the house on our regular visits to Lebanon; during the school vacations of Christmas, Easter and the summer holidays. Most of the hard construction work was completed by the summer of 2006. Six years later, in June 2012, the soft work or final finishing work was completed; Vicky and I relocated to Monsef.

December 2006 our family celebrated Christmas at our new almost completed house; Vicky, Mira, Omar and I came from Kuwait and Majed came from the USA. The house was officially renamed 'Our Home'. Our Christmas present from Mira and Omar was a photo-book titled "*It was a house, we made it a home*". The introduction written by Mira is our reward for the values that hold us together as a family:

Ahlan wa sahlan, Bienvenue, Iervetuloa, Hos geldiniz, Welcome ... to your home.

Your home reflects your personalities...warm, beautiful and comfortable. It is always open for others to enjoy and all are encouraged to relax and stay a while. Your possessions whisper tales of your passions, travels and the journey you have wandered together.
The unmasking of its beauty has been an adventure fraught with water damage, faulty electricity and colorful construction workers. It has taken many years to unfold, after all why do something once when it can be enjoyed 3 or 4 times at least. So, after tiling and retiling, hanging and rehanging, wiring and rewiring...it's time to take a break, have a seat, put your feet up, and stay a while. Can I get you something to drink?

The 'Our' in 'Our Home' refers to the three components making our family: Vicky & Walid; Mira, Omar & Kiira; and, Maria, Majed & Kaila. On the first floor, each component has two assigned connecting rooms. Despite the strong objections of both Mira and Majed and their well-rehearsed arguments, my two rooms are larger and have a nicer view; parental respect and acknowledgment of the source of financing subdued their protests.

The house was originally a single floor: one large space of three cross-vault arcades, used as living and sleeping quarters, and a small single vault room for the storage of the olive oil clay urns. Over many years and three generations of occupants, several bedrooms of different areas and different heights were built over the arcade to accommodate grandparents, parents and children.

Converting this complex structure into a homogeneous villa, functional and with the best possible space utilization was not a small feat; it took several architectural designs and re-designs and the unlimited patience of Aziz, Vicky's cousin, who undertook this great transformation. The end product was a nice looking villa with a non-exceptional external architecture but with two interesting features; the old stone arcade that comes as an unexpected but pleasing surprise when you first enter, a strong contrast with the outer modern façade, and the oil storage room which was converted into the guest bathroom.

The taxi driver helped my father up the stairs, who by then was in his early eighties and experiencing difficulties mounting the few steps on his own, up to the front door landing. Suddenly, the driver lets go of my father, rushes past me to the entrance of the arcade and asks, with naïve curiosity: "How did you manage to move this huge arcade into this villa?" I could have given him the same answer that Ramzi, my sister's huge godson, gave the soldier at the military checkpoint when asked how he fit in the back of the Volkswagen beetle: "I was put in when I was little and now I grew up and cannot get out." I refrained, believing that the Taxi driver would miss the humor in my answer the same way the soldier had; however, with maybe a more peaceful reaction.

I have told the story of Elias and our guest bathroom so many times; I don't believe one more time could make it more boring or less funny. Elias, a relative of Vicky and a good friend of mine, and I occasionally got together, relaxed and discussed the engineering profession over a tumbler of superior quality cognac. Half through our second drink, on his first visit after our house became a home, nature called; he headed to the guest bathroom and I went upstairs. Ten minutes later Elias was still inside with no noise or sound; a slightly worrying situation regardless of the extent of nature's call. "Come in", he responded to my knock on the door; a weird invitation to say the least. There he was, fully dressed and decent, seated with the toilet cover down, reading Mad-Magazine and sipping his cognac. "Grab a chair and join me in this small beautiful and cozy room"; notwithstanding his good intentions, I declined the invitation preferring the more spacious living room. Eventually, Elias bought a large piece of land in Monsef and is currently building a very large and spacious villa; however, it lacks the charm of a built-in cross-volt arcade.

CHAPTER X: 2012 to 2015

"What do you do all day long and every day in Monsef? Do you have friends; do you meet people; are you bored?" These questions are not less naïve than when I was asked during my engineering training year at Hayward Tyler, England, mind you quite a few years earlier: "in Lebanon, where do you park your camel when you go to the university?"

The long years of internal troubles in Lebanon and the subsequent changes in the demographics of cities and towns resulted in a gradual expansion of the residential, business and entertainment sectors along the northern coast; mostly in Jounieh, Jbeil and their surrounding towns and villages. Monsef, having remained as a pure residential area with no commercial or business establishments, notwithstanding its revived school, Fadi's small grocery store and Elia's even smaller primitive bakery for the weekend 'Manakeesh' and 'Corban bread', witnessed an influx in its families back from Beirut and other major cities. The maintained vast areas of greenery and the non-polluted land and air, a rarity these days, have invited the young generation of Monsefites as well as friends and relatives to establish primary or secondary residences in Monsef. Old houses were renovated, fitted with red roof tiles and surrounded by well-manicured gardens. Unfortunately, with all this modernism, the community interaction, other than socializing at homes, was diminished to those who regularly attend Sunday mass, at the church front yard or in the newly built church community hall, initiated and mainly financed by the Sadaka family in memory of a benevolent member, and designed by Adonis.

Adonis

The new church hall is almost the last contribution of Adonis to Monsefites in particular and to life in general; it preserves his memory as a vibrant, generous and active member of any community he happened to be part of. Adonis passed away not long after his Godson; my son Marwan. At the time I was on my second mission in Istanbul. He suffered for some years from an illness that I still do not know its specifics; all I know is that during his last year with us he was admitted to hospital every few weeks to receive a day-long chemical treatment, introduced intravenously. More than once, when he was down and depressed, I would travel from Istanbul to spend those twenty-four hours with him remembering the past and shying away from the future. Adonis was my best friend from the days that we were small boys roaming the valleys and hills of Qornet Al-Roum, hunting birds, and enjoying our camaraderie. As I mentioned, we were called "Sarah and Maynet' in reference to two legendry women who shared the same intimate friendship; I have no idea who they were or what their story was, but it must have been a good one. His days of physical presence with us came to an untimely end while I was still in Istanbul. He remains, through our shared experiences and events, my best friend; unfortunately, his once very close family has, over time, become almost strangers. His wife is now remarried and his two lovely daughters are university graduates and members of the working community. Mayda, his eldest and Vicky's Goddaughter, was recently married; the occasion brought us a little bit closer.

Monthly Trips to Kuwait

Relocating to Monsef was intended to be a semi-retirement. I entered into a consultancy agreement with HEISCO to provide support and advice to the

Contracts Department, for 3 to 4 hours a day, through an internet link between my office in Kuwait and my home-office in Monsef. The rest of the day was supposedly availed for relaxation and for fixing and re-fixing things around the house; a hobby and a passion that has stayed with me since boyhood. With power and non-power tools that I have collected over the years and with my ever improving manual skills, I became the official solution provider to family and friends on all matters related to non-functioning appliances and the one and only approved handyman for hanging paintings on walls. My drilling and hammering skills were definitely improved over the years while helping in the transformation of "our house" to "our home". The evenings and nights, except for those few days where Vicky and I played a one-on-one 'Bereeba', the card game that could still be played well when most of your brain cells are dormant, were intended for socializing at home or at the homes of friends and relatives; mostly over dinner and occasionally over a drink and a Bereeba tournament.

However, as the old Arabic saying goes, 'the wind blows not in the direction desired by the sailing ship.' My working hours got longer and longer until I was almost a full-timer; support and advice developed into full administering of the Contracts Department; and my monthly one-week trip to Kuwait became a necessity, with a heavy schedule, rather than an optional service. Simultaneously, our social calendar got busier and busier leaving very few evenings and weekends for relaxation and mental idleness; that is what I do in Monsef and I definitely am not bored.

My stay in Kuwait during these monthly trips followed a routine broken only when Vicky, once every six months, accompanied me to maintain her residency in

Kuwait and then for our onward trip to Dubai. Vicky's stay in Kuwait, though short calendar wise, is packed with shopping sprees; the money spent is not an issue, since the money you do not spend is money you do not own, but the "pick me up, wait for me and do not lose your temper and start grumbling" over-burden my already heavy schedule. Despite all this, I have a joyful time when she is with me during these trips to Kuwait; this statement is genuinely true and not inserted in this book just to please Vicky when she reads it. The icing on the cake is the trip that follows to Dubai; Vicky to see our children, Mira and Majed, and I, giving her space to spend quality time with them, monopolize the free time of Kiira and the waking time of Kaila.

Is Kiira jealous of the attention I have started giving Kaila, now that she is 7 months old and has started to acknowledge my presence with a smile? I don't think so. Kiira knows quite well that the spell she cast on her "Jiddo", my Arabic title for Grandfather, when she wrapped her 2 months-old fingers around his thumb and gazed into his eyes before floating into her afternoon nap, is irrevocable, unconditional and grants her any reasonable request upon first demand. A reasonable request is when she asks to talk to me in person rather than over the phone; I fly to Dubai the following day for a 24-hour visit. Another reasonable request, though indirect, is when exasperated with the very slow process of earning money by doing chores around the house, imagine only one coin for fixing her bed, she asks Mira: "When is Jiddo coming?" To Kiira, that is the quickest way for buying a new bed that she "absolutely must have"; the spell worked its magic and she got her bed, plus a non-expected but always expected present, after one week from the innocent question. I also enjoyed my 2-day visit to Dubai.

Kaila's 1st Birthday

A recent visit to Dubai, with Vicky accompanying me, was to celebrate Kaila's 1st Birthday; on the 7th of February of the year 2015.

The Birthday Party was held at a public park; a large, green, landscaped, dotted with several playgrounds, housing an exhibition area, would have been crowded if not so spacious, and, most of all, taken for granted as one of many such parks in Dubai. The will to live turned a God created desert into a heaven that people can visit, revisit and come back to tell of the wonders of Dubai. Many of the same people visit Lebanon, some revisit, but all wonder how the irresponsible acts of its people turned a heaven into a desert; a space devoid of most aspects of modern civilization and rapidly replacing its greenery with the dull color of concrete. How can humans manage to develop a long future from a short history and similar humans, at least by Creation, mismanage a long history to achieve an underdeveloped short future?

Deterioration of the Lebanese Population

During the recent past in the life of Lebanon, an infinitesimal part of its very long history, the Lebanese have managed to erode that which took many generations to create: the human element. Internal wars, invasion by outside forces from all parts of the known World, repeated occupation by successive dynasties, World Wars, famine, epidemics, and natural disasters crushed the people who lived in the geographic area now called Lebanon; but the spirit and the soul remained free. Resurrection was inevitable, rebuilding became a habit, and the after was always an improvement on the before. The will to live was a cosmic energy instilled in every man, woman and child. The Lebanon as we know it now is not yet a century old; a republic that got its borders defined, its independence granted after a national struggle, and its constitution written and adopted almost 75 years ago.

Ecstasy was so overwhelming that the Lebanese were blind to the black hole created by the Addendum to the Constitution; a document that was supposed to ensure the rights of all constituting religions and enhance the coexistence of their follower. Instead, it laid the corner stone of the destruction of the Lebanese human element. Religious affiliation shattered national unity.

Without the bond of national unity, the self-discipline needed to elevate the interest of the whole above that of the individual deteriorated to almost zilch. Greed overtook benevolence; selfishness replaced generosity; and the "I" dominated. Corruption, bribery, and the sense of lawlessness prevailed. Despite all this, Lebanon remains one unit; its parts, particles, and components are held together by a willpower that defies explanation. The Lebanese currency, despite the bad economy and the alarmingly increasing national debt, has outperformed all known currencies. The entertainment industry, despite the deteriorating infrastructure and the lack of reliable power and water, has flourished; hardly a vacant seat can be found on any of the very many flights bound to Beirut. The threat of the regional wars cascading over Lebanon and causing a revival of the internal fighting among the various Lebanese factions, the ideological and political opposing groups, and the ever growing religious fanaticism notwithstanding, Lebanon remains a haven for freedom of expression. Revolutions, wars, immigration, expeditions, inventions, knowledge, art, literature, mathematics, sciences and even religions aim at freeing the soul, the mind and the body from their constraints; Lebanon apparently offers the same opportunity for freedom to all its seekers in the surrounding regions.

The erosion of the Lebanese human element is manifested in the complete lack of discipline, an escalating disregard for laws and regulations, an alarming disrespect for family values and, most alarming, disintegration of cohesion of the Lebanese population. I am convinced that anything that has not happened so far, whether right or wrong, is only because no one has yet thought of or decided to do.

Qayseeyee & Samir's Death

Public employees, elected, selected or hired, have not yet felt the urge to fix the potholes in the roads, the shortage of potable water despite the abundance of rainwater, the lack of electric power needed to render long and cold nights bearable, the exorbitant cost of medical care, and the very many defects and deficiencies in the infrastructure and the services that are normally provided to the citizens; so they are not done

On the other hand, Kidnapping, cars with explosives as passengers, snatching of purses by fast-moving motorcycles, cars disappearing to reappear as spare parts or as gifts to non-registered owners or resold to their original owners at discounted prices, driving opposite to the direction of traffic on highways, and very many wrongdoings, have and are still happening because citizens or non-citizens felt the urge to do them.

Someone thought of destroying an important heritage of Monsef: our Qayseeyee; and it was damaged beyond repair. The thousands-of-years-old rocks surrounding the creek were covered by leveling concrete, high enough to render the Qayseeyee a deep hole accessible by unnatural concrete stairs, and visible only if you cantilever your body over the edge with the right balance to avoid falling in it.

With the destruction of Qayseeyee, a major part of our pleasant memories is gone. I remember Samir Banna during his endless efforts to swim, unaided, across the width of Qayseeyee. These memories have been replaced by the sorrow and grief he left behind after the sudden and abrupt end of his train ride through life. I had last seen him when on an impulse he flew from Qatar for a weekend visit to Kuwait. Few weeks later, I was told that he had gone to England for a medical checkup after suffering from headaches and chest pain; he was diagnosed with lung cancer and was gone shortly thereafter. I don't know if our lengthy and frequent telephone conversations did anything to ease his transition from life to afterlife; I know I did not feel any better.

Easter Monday

Today is Good Friday of the year 2015; it is the day to remember Christ's death for the salvation of mankind. The Western Church, following a different calendar than the Eastern Church, has already celebrated Good Friday last Friday and the resurrection of Christ the following Sunday. Religions, supposedly the unifying forces for human beings, are broken into sects that resist unification. Civil wars are raging in the Middle East between Sects; people are being slaughtered because they happened to be born in a religion not popular among the rebels; the election of a president in Lebanon, who by virtue of the Addendum to the Constitution should be a Christian Maronite, is being endlessly postponed because the Christian Maronites of Lebanon cannot agree on a candidate; and governments of the

Middle East region, under the pretext of protecting one religion or another, are unanimously committed to the support of every possible effort that enhances local and regional fighting and killing of their peoples in order to save their peoples; is this a paradox or not! The atrocities committed under the auspices of religions defy every doctrine of every religion. Dhour El Shwair, a town in Lebanon populated by Christian Orthodox and Catholics, followers of the Eastern and the Western Church, disregarded all norms of religious disparities by celebrating the resurrection of Christ on one and the same day; no more Easter and "Wester" in Dhour El Shwair.

The religious rituals, celebrations and festivities of Easter start mainly on a Thursday, when Christ was crucified, and end on Easter Monday, the day Christ appeared to his disciples and proved his resurrection. Easter Monday is also the religious day sponsored by the Nasr family; starting with my grandfather, or maybe before, and passed on to me through the traditional chain of elder sons.

This concept of sponsoring religious holidays is probably unique to Monsef. It started way back when very few families were blessed with the means to eat well, dress nicely when the occasion called for it, and send their children to schools outside Monsef. Yet, these few families remained an intricate part of their community and were committed, out of loyalty rather than obligation, to share these material privileges with the rest of the Monsefites on those days that demanded joy and happiness; they are the religious days that, though not many, were scattered over the year and were enough to re-kindle the flame of joy.

Years ago, when I was still a child, the festivities of Easter Monday at my parents' house started noon time, just after the end of Mass, and lasted till early morning of the following day. It was an open house for family, relatives and friends, where an invitation was not required, and the only obligation was to eat, drink, sing, dance and be merry. Those were the days when family ties were strong, loyalty was a given, respect was granted not demanded and social unity was an unbreakable bond. With the advent of technologies, the increase in the size of families and family obligations, the move of people out of their local community and into international communities and the pressures imposed by growth and development have all helped to erode this social bond. This Easter Monday our guests, by invitation and not by a social trend, are very close family members who will join us for food and drink; a 3 to 4 hours gathering of around sixty people who are hoping for the rain to stop so that the children can have their egg-hunt in the garden and the grownups can smoke freely on the front patio.

Getting Physically Old

When's a good time to get off the train? Some decide to jump off before reaching their normal destination. Others cling to their seats and prefer to be carried off when their self-propelling system fails completely. However, most of the passengers of life prefer to leave their alighting to the will of forces beyond their control; usually, when the human organs decide to disobey their harmonious interaction and the body succumbs to disease or deterioration by old age.

Along the train ride some passengers live through a transitional period; when the aging of their souls does not keep pace with the aging of their bodies. This disparity between feeling young and actually being young is more flagrant after the joy of bouncing on the dance floor for few hours wears off, and the need for muscle relaxant and headache medicine is a prerequisite for few hours of sleep. Hosting Easter Monday lunch is another occasion to envy those who are mentally and physically young. This realization is arrived at by experience rather than wisdom or psychic reflection; Vicky and I are good references for validation. Luckily we have Dr. George, the pediatrician of our children and grandchildren who, out of friendship, extends his practice to family medicine when the high spirit settles down and the body resumes its aging process. Dr. George has recently extended his free-of-charge service to providing our Saturday breakfast of "Manakeesh'; the thyme and oil covered bread, prepared by his mother and cooked at his uncle's bakery.

There are long-riding train passengers who are destined to the misfortune of unequal or non-parallel aging of the body and the mind; either a healthy but abandoned body with no controller or an alert mind that is frustrated by a non-responding mass of flesh, bones and muscles. In both cases, it is a disobedience that ends with the capitulating of the mind and the failure of all body and mind supporting organs.

July 12, 2015

We have been married for forty years; four decades packed with actions, activities, incidents and accidents all made easy and bearable by the love we nourished within our family. Dwelling in the past is a refuge from the present; thanks to the ability of the mind to filter out sad moments and remember the happy moments or those conceived to be happy. The ride on the sledge, drawn by a horse and not a deer, during a snow storm 30 years ago, with the family huddled together and totally covered in thick sheepskin, from Megeve to a remote restaurant high in the

Alps to eat couscous, is remembered with a smile. The recent turmoil in the Middle East, culminating in the war in Syria with bizarre and abnormal growth of influence of ISIS, which brought unforgivable barbaric destruction of human life and the remnants of the oldest known civilization, shames the present and the foreseeable future.

Nine years ago, to the day, Israel launched a massive attack on the infrastructure of Lebanon, destroying almost every bridge and vital road, as a punishment to a country that does not stop Hezbullah from being a potential nuisance to Israel; the attack, that resulted in destruction beyond Lebanon's financial means to rebuild, was in retaliation for kidnapping two Israelis by the militants of Hezbullah. We learned in Physics that every action has a reaction; this theory does not accommodate an action to reaction ratio of one to a zillion.

Ten days ago I became a grandfather for the third time. Maria, thanks to Majed's few minutes of participation nine months earlier, gave birth to Alexander; a 3 kg and 47 cm baby (obviously a boy since he was named Alexander) blessed with chubby cheeks and insatiable appetite. When not sleeping, feeding is demanded; breast or bottle is equally accepted with pleasure.

December 14, 2015

Sixty-seven years ago, somewhere between the flight time of Cinderella in her pumpkin carriage and the roosters waking call, I was delivered to the train of life to join billions of human passengers; each on his or her trip to somewhere. In Monsef, it is the sixth hour of the fourteenth day of December, the last month of the year 2015. Vicky and I are already having our morning coffee; on no-action evenings we tend to observe the rule of rural dwellers: "early to bed, early to rise, makes a man healthy, wealthy and wise". In truth, I do not believe this is helping much; our consumption of health supplements and medication for high-blood pressure, Cholesterol, Glucose, HDLD, LDL-C is maintained at constant high cost to our Insurance Company.

In Dubai, where I would be older by two hours, Kiira is already in school mastering the intrigues of second grade. In England, Majed is probably still asleep; the first week of his two-week business trip has been successfully completed. At other places around the World, billion others are doing what they do at whatever hour it is within their time-zone. Also, millions of these billions are preparing for the celebrations of Christmas and the New Year; some to give for the joy or the tradition of giving, others to receive for the satisfaction of receiving and many more in need to receive to momentarily ease the hardships of survival. Of the last category, the custodians of human misery, the numbers are increasing at alarming levels; worldwide in general and in the Middle East in particular. The updated millennium UNDP report for most of the Arab countries could show a decline in birth rate and an increase in mortality rate. The former due to abnormal escalation of cost of living and the latter due to disappearance of sources of reasonable income; both brought about by the raging wars and the residual political and social conflicts. It is called "the Arab's spring time". What an insult to human intelligence and a stab-in-the-back of the Arab's history. This is not a social revolt against suppressive political regimes or a sudden thirst for freedom and self-expression. It is a human volcano that matured, erupted, and soon would calm down, leaving nothing but destruction all around it, waiting for it to re-build its internal inferno until it erupts again and again; a cyclic plague that has inflicted this part of the World for centuries.

Lebanon is still a republic, a nation and a sovereign country; in theory rather than in practice. The position of President of the Republic, the head of the nation, has been vacant for almost two years. Hopefully, all the political forces in the East, West and in between would soon decide that it is time to appoint a President and go a step further and decide on the lucky winner. The details would be left to the Parliament to work out. However, the Parliament term expired few months back. The members unanimously approved an extension of their employment, without the need to consult their employer, the people of Lebanon. Pending agreement on a new electoral system to be followed by election of new members or the same members or their heirs; alternatively, a judicial decision revoking the unilateral extension of employment; or, the employer terminating the employment; the legality of the Parliament's decisions is questioned and legislative paralysis remains. The Cabinet is another story. Unless a President is appointed, the current Ministers cannot enact new laws; they restrict their efforts to the day-to-day activities of presumably the administration of the Country. Unfortunately, these activities are mostly restricted to interaction with the media and a race to locations of disaster and conflict for a show of not-so-true sympathy and to offer empty support. Providing reliable electric power; reasonable supply of potable water; barely adequate healthcare; collection of garbage; treatment of sewage; maintenance of infrastructure; creation of job opportunities; enforcing the law; relief of the 24/7 traffic congestion; ensuring minimum levels of security and safety; protection of the environment and prevention of eroding what mother nature took thousands of years to create, are all on hold "waiting for Godot".

Despite all of the above, and a lot more social and political deficiencies, every flight to Lebanon from whatever destination is almost fully booked. Every restaurant, regardless of the exorbitant prices, requires reservation in advance at every meal of the day. Famous international entertainers, singers, musicians, actors and performers, individuals or groups, are overjoyed when invited to perform in Lebanon to full houses; tickets are always sold in advance. A young fellow traveler on a flight from UAE to Beirut, occupying the seat next to Vicky, sighed with utmost pleasure when the plane touched down at Beirut Airport. His response to Vicky's questioning look: "Garbage or no garbage, there is no place on Earth as good as Lebanon." The unconditional loyalty of the Lebanese to Lebanon is boundless, unlimited and is passed on from generation to generation, even among immigrants of long time ago. The descendants of these immigrants have been granted recently the right to acquire the Lebanese citizenship; a status that was forbidden to their forbearers. Hopefully the seekers of the Lebanese nationality would soon balance the runners away from it. With the continuing conflicts around Lebanon and the influx of refugees, the ratio of non-Lebanese to Lebanese nationals, in Lebanon, is increasing at a rate that threatens the delicate equilibrium among the various factions; religious as well as political.

Today, tomorrow and one dozen after-tomorrows shall pass before the train reaches the "Thank you for visiting Year 2015" sign and, few puffs later, the "Welcome to Year 2016" sign. These couple of weeks distinguish themselves with long eating hours, short sleeping hours, the giving of many gifts, the receiving of much less gifts, praying, partying, receiving friends, visiting relatives, contacting the end-of-year-remembered-acquaintances and enjoying family get-togethers. In

Lebanon, this is more than a well-kept tradition; it is the annual renewal fee of the best insurance system in the World: The Family.

In addition to meeting all the customary and traditional activities of the Christmas Holiday, twice and thrice over, our family celebrated the Baptism of Alexander; a Christian religious ceremony that Abouna Emile, Monsef's priest, insists on concluding with three quick dips in an urn full of prayed-upon oily water, despite the strong objections of Maria and Majed and the screaming of Alexander, mostly in surprise rather than lack of oxygen. The occasion called for a heavy lunch for the attendees, restricted to direct family members of Maria and Majed, followed by a short nap and then back to the routine of Kaila and Alexander; playing, smiling, crying, drinking milk, changing diapers and going to sleep. This lasted a bit longer than usual, despite the help of Kiira, as a special treat for Vicky and I; Mira, Majed and their families were going back to Dubai a day before New Year Eve to meet the Year 2016 in peace, tranquility, fair weather and the opportunity to be optimistic for another happy year.

Appendix I – Management Extracts

P33
I have always practiced self-criticism, self-evaluation and self-assessment in my belief that these lead to self-improvement, and always came to the same conclusion that fear of failure was not caused by my insecurity but by the dire need to achieve and to be recognized for my achievement. The importance of expansion of knowledge, the urge to achieve and the significance of reward are basic principles that I adhered to throughout my management practice. I even encouraged others to seek and accept criticism; however, that was something that I preached but failed to practice.

P35
Table Tennis was instrumental to maintaining and improving my sharp reflexes, which through my career days proved to be a very important management tool when dealing with people or ideas alike.

P36
Make good connections; offer and give help and assistance when you can without asking for compensation; be and not act sincere about helping others; and be thankful when people return your favors. As long as there is no material or money involved in asking and receiving favors, and as long as you stay noticeably ahead in the exchange of favors, most of the people who possess sound human values would be glad to reciprocate.

P39
The message was delivered, without a word being said, that the world does not run by my rules or per my expectations and that it pays off to be patient rather than jump to conclusions, especially if having a poker-face is not in my character.

P39
It was at Hayward Tyler where I realized the significance of Trainer-Trainee relationship and the importance of "Transfer of Knowledge" in order to contribute to the development of the human element and to the community in which we live.

P45
Always learn about the cultural, social, and religious practices of any community before you move into and interface with.

P47
Humor is often wasted on government employees, especially if they take their job with more seriousness than is warranted.

P47
To participate intelligently in the exercise of assigning what piece of equipment to do what job at what construction site, for a most efficient and cost effective utilization.

P49
He had given the store keeper a wrong part number for a defective oil seal, a small error that in the construction industry could be the cause of costly delays, wasted manpower time, and maybe contractual difficulties.

P50
Take it easy for the rest of the day; think about the effect that George's immediate departure would have on your work; assess the impact of losing your best mechanic in the Company; see if you can salvage the damage in your relationship with your employee without losing your authority; and in the morning take the action that you deem best.

P51
Management is an art that you are born with a flair for; a skill that had to be continuously polished; a science that needs a lot of studying; a complex issue as diversified as life itself; and a life built on experiences, with each adding a new lesson to be learnt and stored in an active file. In short, management is as complicated as its main and primary elements, the human being. No two persons are alike; no two people react similarly to the same stimulus; no individual reacts in the same manner to different acts; and it is rare that the same person reacts in the same way to the same incident at two different times. I knew that I possessed the basics for a career in management, but had a long way to go to develop it. The ease of my interaction with others, whether friends, relatives or complete strangers, and regardless of age, education, profession, gender or nationality, was

so natural that it quickly dissolved any behavioral barriers which hinder smooth communication. Courage to be outspoken and to voice my opinion was something that I never lacked. Some might call it being argumentative, opinionated, or even annoying; however, I always managed to articulate and to defend my point of view. From my early childhood, I was raised by a family that did not believe in classifying people by race, gender, the so called social status, or any other factor that puts an individual on a higher pedestal than others. People do differ by factors beyond their control such as in-born characteristics, intelligence, artistic temperament, patience and rich vocal cords; by genetically inherited physical features that result in blue eyes, dark hair, or muscular structure; or by naturally occurring alternatives resulting in being one of the two genders, or with dark or fair skin-colors. These factors should not be conducive to preferential treatment. What distinguishes individuals is the will to learn; personal achievement; aspiration for knowledge; perseverance when faced with difficulties and obstacles; loyalty and compassion to one's ideals; and most of all the courage to stand up for one's ideas and beliefs. When acknowledging the inevitability of these differences and the reverence of honest achievement, interrelationships are reduced to acceptance of the former and respect for the latter, and individuals are regarded as humans and not holy beings.

P52
To manage you need to maintain sustainable communication with others.

P56
At that very early stage in my career I decided to adopt the same interview technique, as interviewer or interviewee: meet the other party and get to know the quality of the person before getting into work records; and treat the personal work resume' only as a document for screening purposes and selection of candidates for interview. Success and achievement are very much affected by work environment and opportunities. With the right personal attitude and positive character traits, a non-achiever could turn out to be a significant producer when properly directed and well managed.

P57
Caring for the welfare and the professional development of your employee is one thing, and assuming the right to dictate their personal life is another; the latter could never succeed as it infringes on the right of personal freedom.

P58
In order to perform well in the engineering business, whether design, consulting, or construction, one does not have to reinvent the wheel. The secret was to know what information you need; where to look for it; how to find it; how to select the most suitable data for the required application; how to integrate it in the overall solution; and, most of all, how to do all that efficiently, on time, and at the best cost.

P59
I sometimes felt that topics were raised for my benefit; later on I became aware that it was "hands-on- training" through daily interaction.

P71
Construction Methodology Department whose function was: The evaluation of design criteria in conjunction with the latest modern construction philosophies and the most advanced material and equipment with an aim to achieve a most economical execution plan.

P74
My power from then on was an extension of me only, coming from my knowledge, capabilities, and experience.

P78
The Safety Manual recommended that traditional local practices be reviewed and evaluated in comparison with international adopted procedures, and if found safe for use for the intended purpose they could be applied under supervision of someone well versed in these local practices.

P81
Liquidated damages, sometimes known as penalties, represent financial compensation to the client for the client's suffering as a consequence of not being able to use the facility, as intended in the contract, on the date specified therein. My responsibility, as the project manager, was to prove that the delay, which was a fact, was not caused by Envirogenics. There are two ways of doing that: either

to prove that the delay is fully attributed to the client, in that case the Royal Navy or the Corps as its representative, or that there were concurrent delays caused by the client and Envirogenics. With concurrent delays, as long as the contract has one completion date for the full scope and not interim partial dates for parts of the project, the contractor can argue that his delay was due to his rescheduling of his work to span over the new period extended by the client's delays. Arguing is one thing and winning the argument is another. For a start, an airtight case has to be built with enough documentation to address every possible counter-argument; updated work schedules with activities logically sequenced and properly linked; supporting correspondence and minutes of meetings sorted in chronological order; and a brief of contract terms and conditions that support and justify the claim for non-responsibility for the delay and thus waiver of any penalties.

The importance of cross referencing and tracking of documents was first instilled in me at IC, my high-school, during my tenure as chief-of-stacks of the school library. Documenting all issues related to a project, including discussions, instructions, and suggestions, by means of correspondence, minutes of meetings, and official reports, is an important management tool that was advocated by Bechtel; a technique I fully embraced and reaped its benefits on more than one occasion.

P83
Half the battle is won if you believe in what you are fighting for.

P86
The difficult part is to first convince myself and then convincing others becomes much easier. I present an argument to myself, supported with facts, and try to shoot holes in it. Weak or unsubstantiated facts are eliminated and the holes are filled with more reliable ones. Substantiating documents are collected and the cycle of argument versus counter argument is repeated. With each cycle the case gets stronger, better structured, and the facts well linked and streamlined. When my belief in the righteousness of the case becomes unshakable, I present it in a straightforward and focused manner with a sharp eye kept on the facial

expressions of the members of the other party to gauge the level and the extent of converting them.

P87
The Initial Acceptance Date is the first of the two completion dates that are usually set in a construction contract. It is the date on which the Client, the owner of the project, can start using the facilities of the project as intended, even though the scope of work of the contract is not completed. Uncompleted work could be work that was either not done, or was done but is not in accordance with the technical specifications of the contract, or has some deficiencies that need to be corrected.

P87
The Client has the final say as to when he or she is willing to occupy that room in spite of uncompleted work. Of course, the Contractor is not relieved of his responsibility to complete the outstanding works; however, the Contractor would not be responsible for damages caused by the Client while occupying that room. That is called beneficial use by the owner of part of the project before the works are contractually completed. The second completion date is the Final Acceptance Date, the date on which the Contractor is considered having fulfilled his obligations under the contract, including the completion of outstanding work identified at the Initial Acceptance Date as well as rectifying any defects that appear between those two dates. This period between Initial and Final Acceptance dates is usually called warranty period, during which the Contractor remains legally and financially obligated towards the Client for any uncompleted or defective or deficient work. To ensure that the Contractor would honor his obligations after Initial Acceptance, the Client holds a percentage of the Contractor's earned money as Retention Money and would maintain the Performance Guarantee issued by the latter in favor of the former.

P96
A Director of a company called in his Engineering Manager and asked him how much is 2 plus 2. Suspecting some deep connotation for this simple question he spent two hours researching through his technical library, but the only answer he could find was 4. When the Administration Manager was asked the same

question, and after hours of reviewing all the company manuals and procedures, he gave the same answer of 4. The Financial Manager took a full week to come up with the same answer. However, when the same question was put to the Legal Manager his response was instantaneous: how much would you like the sum to be?

P96
Upfront payment in the contracting business is a temporary relief of a company's financial difficulties and not a cure. It is a sword with two edges. Unless the root issues are properly managed and resolved, the company would eventually succumb to these financial burdens; the answer of the Financial Manager would be more befitting than that of the Legal Manager.

P97
The variance between progress based on billing and progress based on cost was well understood by Sogex management. It starts with a scheduling and work planning exercise which, when approved by the parties to any project, produces the work program that monitors the execution of works, the ordering of material, the manpower allocation, and the payment plan of that project. First, a list is made of all the activities that need to be accomplished to execute the project, from start to completion. These activities would include engineering and documentation; material procurement and delivery; mobilization to site and demobilization; establishing of offices, workshops, and other on-site facilities; execution of work; and testing and commissioning of the facility. Then these activities are linked in a proper sequence within a logical network. You cannot paint a room before you buy the paint, and you need to excavate a trench before you can backfill it. A time estimate, or duration, is assigned to each activity by experienced people in the related field of work, who usually have a databank of related information. A lot of this information is published by authorities in the construction industry; however, each company uses data based on its past performance. The activities are then resourced; manpower is assigned by profession and number of working days, material for each activity is identified, and the necessary equipment is reflected. Based on established costs of these resources, each activity would develop a price. Using any of the commercially available computer software programs that are designed for project planning, the durations and resources of the activities are added to arrive at a final project cost

and a date for completing its scope of work. Since a project usually has a restricted time period for its completion; the contractor has a pre-estimated budgeted cost for the works; and manpower resources are limited, the network logic, the durations of the activities and the resources are modified several times before arriving at a satisfactory program with optimum resources utilization and cost estimates. Logic, experience, construction management skills, and cost awareness are essential characteristics of a planning and scheduling expert.

P98
A well-resourced work schedule with proper costing of its activities would reflect overall percentage completion relative to the total incurred cost against these activities. However, since contractors tend to be secretive about their actual costs, mainly to keep their profits unknown to other parties; since it is much easier for clients and contractors to measure progress by directly estimating the physical completion status of an activity; activities are given monetary values, mostly derived from the contract priced bill of quantities, with the sum of these values adding up to the full contract value. At the end of every interim reporting period the percentage completion of every activity is assessed; these progress percentages are converted to money; and the total sum would be the billing for that period. Overall project progress is the ratio of total billing to the total project cost.

P103
The basic idea is to always choose the tool that best fits the intended or required output; over-specifications are a waste of money and under-specifications result in a product of poor quality.

P115
At the early stages of a construction project, assessing and quantifying the monthly work progress is subject to approximation. The extent of this approximation and whether it is made on the plus side or the negative side depends largely on the relationship between the Client and the Contractor. After all, contract procedures allow interim overpayment or underpayment as long as the measured work is reasonably quantified and the estimated percentage

completion is close to reality. As the project nears completion, quantities of work are more accurately measured; a closer look is taken at the quality of the presumably finished work to identify corrective actions or required replacement; and the Client tries to insure that enough money is withheld from the Contractor's payments to cover for all the outstanding work. It is a common knowledge in the construction industry that the cost incurred to move from 90% completion to 100% completion usually exceeds the remaining money to be billed and received. The definite outstanding receivable sum is fixed and is the difference between the contract value and the actual amounts paid. Potential receivables are claims submitted by the Contractor for what he believes are extra works, for which the Client is still to accept and approve. The cost to complete is elastic; the longer it takes to complete this last ten percent the higher the cost would be, and more so if new defects arise at this last stage. Ideally, the cash plan of the project should accommodate this negative flow at the end of the project, and the Contractor should restrain himself from over interim billing despite the temptation of upfront payment.

P141
Building up an organization to handle such a construction project had to start with identifying and defining the elements involved. These elements fell under the following categories: Engineering & Design, Planning & Scheduling, Material & Procurement, Office Administration & Accounting, Quality Control and Safety Procedures, Quantity Surveying and Cost Control, Offices & Accommodation, Lay-down Areas and Productions Facilities, Manpower, Equipment, and Financing & Bank Facilities.

P142
Site engineering is essential and it mainly involves preparation of shop drawings to translate an engineering drawing into as many detailed sketches as needed, which would then be used by the artisans on site to build from. By this way an element can be broken up into components that relate to different specialties for ease of understanding. Site engineering is also important to develop design alternatives in order to simplify and speed up execution, reduce cost, or deliver improved product without jeopardizing the intended quality of the finished works. Cost savings to a contractor by speeding up execution, simplifying the

construction, and handling the design of additional works, more than pays for a site engineering unit.

P142
If a contractor had no engineering responsibility then the Client would have required the Contractor's staff to be lawyers or doctors rather than professional engineers. It makes sense.

P143
Planning is defining a logical path to reach a pre-set and attainable goal. We plan in order to avoid disruptions, distractions, and possible failure prior to achieving our goal. Planning is a must if we have constraints along this path, such as time and cost. The level of detailing should optimize the number of activities and their complexity. The activities or actions should remain manageable, controllable, continuous, and most importantly one-directional with a focus on the required goal.

P144
Planning is logic; scheduling is knowledge; and compliance with both is self-discipline. Engineering is a good conduit for developing logic and knowledge. Discipline is a personal characteristic trait that most likely has its own gene somewhere along the DNA chain. Unless it is practiced, discipline loses its glitter and eventually fades away. However, it should be practiced in moderation with allowance for spontaneous actions.

P145
Engineers and designers must not restrict contractors to a specific brand for any material item unless absolutely necessary. The contract documents should state material specifications in accordance with the requirements of the project. If guidance is essential, due to limited suppliers or complex details of the item or other compelling reasons, the engineer could mention a brand name and add "or equal". This would prevent design engineers from favoring one manufacturer over

another for personal gain, provides equal opportunity to all suppliers, and gives contractors a wider choice to seek competitive prices and better quality.

P145
All these materials are identified in the contract Bills of Quantities (BOQ), the document that lists all project items, their quantities and prices which, when added up, form the total contract price. Material items are listed as supply and install.

P148
The responsibility for assuring good quality final product, equivalent to what the owner paid for, has shifted from the individual worker to a QA/QC program managed by specialists whose function is to monitor the quality of work, guided by the procedures of the program.

P149
The procedures and regulations are intended for assuring the health, safety, and survival of the people associated with the project. It extends as well to the protection of the works from damage; the safety of equipment used for construction; and the safety and security of the means of access on site.

P150
Contractors are usually allowed a mobilization period at the beginning of the project to provide the necessary resources and facilities for the start of work.

P152
Production, efficiency, and quality, all boil down to the human element. Management sets policies, provides guidance, monitors performance, avails the tools, and hopes that the evaluation and selection criteria for hiring appropriate performers are effective.

P152
Re-assignment and re-re-assignment of workers to different tasks and a continuous on-site training were the means used for "selection of the fittest – most suitable for a task-".

P153
The secret to getting the best out of a person is the fact that every person has something positive to offer. The successful manager, whether he is a foreman, supervisor, manager, or director, is the one who could bring out the best in others in the shortest possible time.

P153
Recruitment is the selection process; training is for development of skills; procedures are for guidance to better performance; and work environment is for improving productivity. Personal treatment is what binds all these together to ensure sustainable growth;

P156
Like every prudent bank, the officer-in-charge reviewed the contract terms and conditions; assessed the political climate; studied thoroughly the project's budget, cash-flow, and cost elements; and assessed the risks of default prior to approving the financial package which consisted of cash loans and bank guarantees.

P158
Arbitration, a form of dispute resolution, is becoming more and more the contractual and legal venue for resolving conflicts between contracting parties. It is intended as a more efficient and much faster means of legally resolving conflicts than going to overburdened courts of law.

P180
The TOR, which is the basis for the proposals, should contain comprehensive information as to the requirements of the beneficiary. It needs to be in a unified format to deliver a clear understanding of these requirements to the consulting firm, and should be linked to the evaluation process to ensure a fair and equitable evaluation of the proposals by the members in the sub-committee. The evaluation criteria should be standard and must follow internationally recognized and

accepted systems to ensure the hiring of the most appropriate consulting firm for any particular project. The bidders should be selected through a process that matches experience and expertise with the scope of work; ensures equitable distribution of projects among local firms based on their work load; allows the choice of international consultants that would provide an added value to the project and the transfer of knowledge to the staff of their local partner; and takes into consideration the past performance on Government projects with respect to quality of work, timely production, and completeness of product. Evaluation of proposals should be carried out by qualified staff, of proven integrity, with each member working independently; individual results are then combined for final scoring. The financial proposals should also be evaluated and their scores added to those of the technical proposals, using a pre-agreed upon weighing system that reflects the beneficiary's preference of quality over cost, or cost over quality; the complexity or simplicity of the project is the main deciding factor.

P194
Organizational philosophy is based on interdependency among functional units that are segregated into two groups: Operations and Services. Project and construction management and execution of works on site are the responsibilities of Operations. Everything else falls under the headings of Support and Services: Business Development to secure clients and projects; Proposals to price and win tenders; Safety and Quality Assurance to ensure healthy and safe work environment; Project Controls to monitor time and cost and advise on corrective measures when the project derails; Contracts to handle all contractual and legal aspects, before and after contract award; Finance to manage revenues, costs and hopefully profits; Human Resources and Administration to take care of the manpower element; Equipment Division to secure the machinery and equipment as required by Operations; and Procurement provides the materials and external services. In short, projects are secured and handed over to Operations with all the necessary tools, guidance, monitoring, control and advice with one obligation: manage the execution of the works to complete on time, to the required quality and safety standards and make profit. There is another organizational system that treats each project as an independent profit center where all the Support Services are contained within the project and are, in addition to execution, under the control of the project manager. Each of these two systems has its own supporters, defenders and advocates.

P195
The Contracts Department is a service provider; as such, and being a newly introduced concept, it was important to promote it within the company with delicacy and diplomacy rendering it a service that is sought after rather than an imposed one. It is the usual balance between leniency and firmness, between flexibility and rigidity and between success and failure. The Contracts Department, in few months after inception, was recognized as a one team of serious and studious professional service providers;

P196
For an adviser, the easy part is giving an advice. The difficult part is getting others to seek your advice; to respect it; to accept it; to benefit from it; not to be threatened by it; to recognize that it is delivered free from any ulterior motives; and to come back for more. It goes without saying that the advice must be founded on solid, multi-facet and diversified experience, and it should be delivered with an aura of maturity and a non-contending manner.

Contracts Management covers but is not limited to contract formation and contract administration. The former deals with all contractual activities prior to concluding a contract agreement, which is a legally binding document, such as: Review of contractual and legal terms of tender documents; preparation of contract agreements (Rental Agreements, Service Agreements, Consultancy Agreements, Agency Agreements, Cooperation Agreements, Joint Venture and Consortium Agreements; Pre-Bid Agreements etc.); developing procedures for contract administration; and developing and maintaining monitoring tools: forms, formats and logs. The latter provides direct support to the project team executing a concluded contract to ensure compliance with all legal and contractual requirements of the contract and all sub-contracts; vets correspondence of contractual nature, an important tool that could work for or against the interests of the project; and handles all claims by or against the project.

Contracts Management is involved, in addition to contracts at the pre and post award stages, with the overall management of a company where such *management directly or indirectly affects the performance of the projects. Company policies and procedures, organization structure, performance of support departments, relations with clients and other business partners, management of company resources (manpower and equipment) and company strategic planning are aspects of the overall company management.*

P197
Each Contract Administrator has two functions: prime function to support the business unit he is assigned to and the secondary function to assist other Contract Administrators whenever the need arises: to cover up for each other during leaves, absences and work overload. The secondary function required a team spirit: continuous communication and cooperation among the Contract Administrators, support and mostly the sharing of experience and expertise. The purpose of CAD monthly meeting is to reinforce the One-Team spirit and to exchange lessons-learned from interesting cases and issues; more importantly, to promote camaraderie.

www.ingramcontent.com/pod-product-compliance
Lightning Source LLC
Chambersburg PA
CBHW021358210526
45463CB00001B/143